I0824755

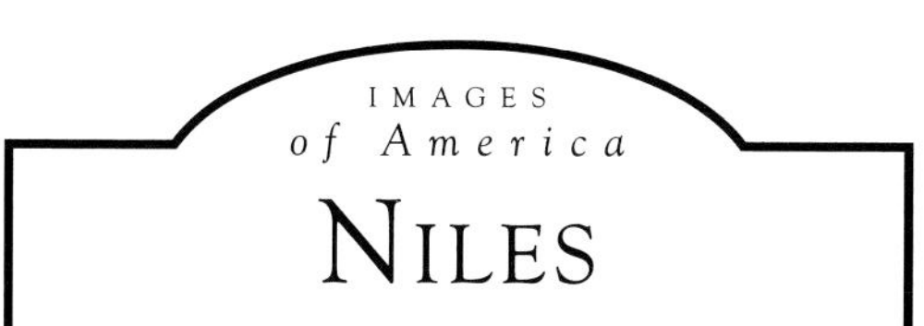
IMAGES
of America
NILES

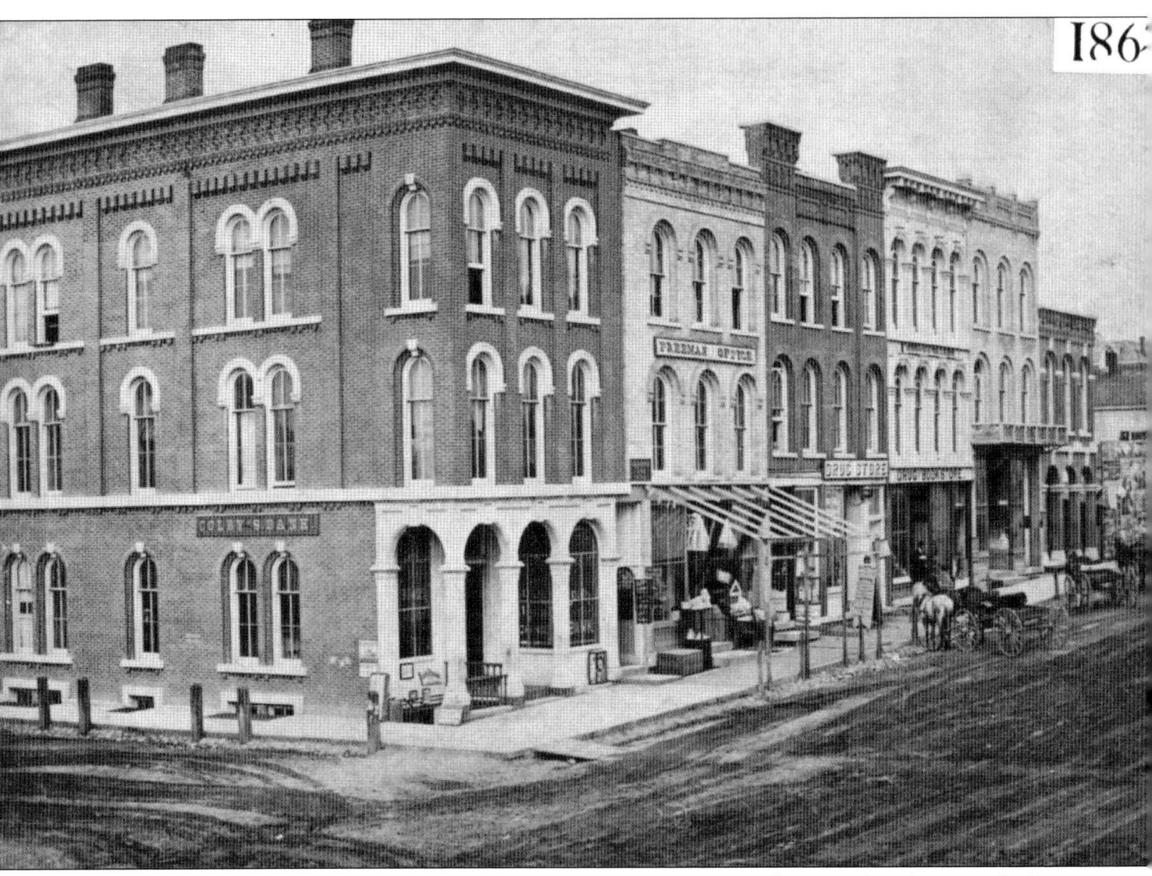

One of the earliest images of downtown in the Niles History Center collection, this photograph shows a view looking east from the corner of Main and Second Streets in the late 1860s, when hitching posts were necessary for patrons to utilize downtown businesses. The northeast corner building dates to the 1850s. Colby's Bank is seen on the lower level. (Courtesy of the Niles History Center.)

On the Cover: The Main Street bridge connected the east and west sides of Niles. This image dates to 1900 and shows carriages crossing the St. Joseph River heading to the fairgrounds on the east side of town. This bridge was constructed in the late 1860s and replaced an earlier wooden bridge. New bridges were constructed in 1920 and 2015. (Courtesy of the Niles History Center.)

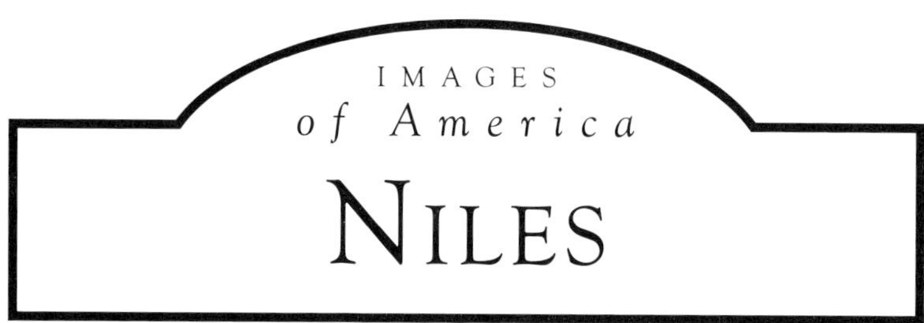

Christina H. Arseneau and Mollie K. Watson
Foreword by Mayor Nick J. Shelton

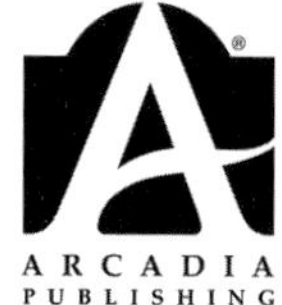

ISBN 978-1-4671-6324-8

Published by Arcadia Publishing
Charleston, South Carolina

Printed in the United States of America

Library of Congress Control Number: 2025948907

For all general information, please contact Arcadia Publishing:
Telephone 843-853-2070
Fax 843-853-0044
E-mail sales@arcadiapublishing.com

Visit us on the Internet at www.arcadiapublishing.com

To all Nilesites, past, present, and future

Contents

FOREWORD

In all 39 years of my life, I have come to know Niles in a way that is hard to put into words. I was born here in 1986 and, like so many in this community, I grew up working alongside my family. For me, that meant long days at Shelton's Farm Market with my dad, Mike; my grandpa, Jimbo; my uncle, Joe; and my cousin, Chad Geister, who we sadly lost this past year. The market closed its doors in 2024 after more than seven decades, but the lessons I learned there about hard work, integrity, and community will stay with me forever.

I have also spent the last decade of my life serving as mayor of Niles. First elected in 2016, this role now makes up a full quarter of my life. I stepped into office at a time when America was hungry for change, and I have been proud to spend these years working alongside so many dedicated people who care deeply about this place.

Niles is where my wife, Allison, and I have chosen to raise our family. We were married in 2017. Allison is a Niles High School graduate, and I am a Brandywine grad. Raising our three children, Emery, Rocky, and Mickey, in the same community that shaped both of us has been the greatest reward.

I have lived in other cities: Las Vegas for a year and Ann Arbor for a year, and both are wonderful in their own ways. But they are not home. Home is the St. Joseph River, Friday night football games, live music at the NODE (Niles Outdoor Downtown Experience) or the amphitheater, neighbors who know your name, and the feeling that you are part of something bigger than yourself. That is Niles. Looking through the images in this book, I see the roots of that feeling. They show the people who built our industries, walked our Main Street, created our schools and churches, and made sure there was always something to celebrate. Their stories live on in the streets we drive today and in the pride we carry forward.

As mayor, a lifelong resident, and now as a parent, I could not be prouder of this city. My hope is that the images in this book remind you not only of where we have been, but also of the community we continue to be, a place that values its past, embraces its present, and believes in its future.

—Mayor Nick J. Shelton

Acknowledgments

We would like to acknowledge the city council of Niles, Mayor Nick J. Shelton, our city administrator Richard Huff, deputy city administrator Joe Ray, and the entire City of Niles staff. Thank you for supporting this project and recognizing the importance of the past in planning for the present and future. Thank you to all of our volunteers and docents. You support our work in many ways: greeting visitors, leading tours, scanning, cataloging, and decorating. To our coworkers Mike Kiernan and Halie Riffett, thank you for keeping the place functioning while we were writing. To the Friends of the Niles District Library and History Center, staff of the Niles District Library, Dowagiac Area History Museum, Pokagon Band of Potawatomi, Support the Fort, Erika Hartley, and the Fort St. Joseph Archaeological Project, thank you for all of your historic work. Thank you to our Arcadia publishing editor, Caroline Vickerson, for your timely feedback and support. Finally, we thank the Greater Niles community. This book is dedicated to you.

All images used in this book are from the Niles History Center unless otherwise noted.

Introduction

Through the lens of photographic collections at the Niles History Center, this book chronicles the people, places, and events that shaped Niles as it grew from a village into a city through the mid-20th century. Niles developed at the crossroads of land and water. At a key river crossing where trails converged, the area became known as Pawating, meaning "place of the rapids." Historical records, drawings, and maps provide a glimpse of the city's foundations when photographs do not exist.

French explorer La Salle scouted the area in the mid-17th century and encountered native people who had lived here for centuries. In the 1680s, Father Jean Claude Allouez established a Jesuit mission along the "River of the Miamis," now called the St. Joseph River. Close to a vital portage connecting the Kankakee, Illinois, and Mississippi Rivers, the site's strategic importance led to the establishment of a permanent post in 1691. Fort St. Joseph became one of the most significant settlements in the western Great Lakes region. For nearly a century, the fort served as a religious mission, trade center, and military post. In the mid-18th century, Fort St. Joseph ranked fourth in fur-trading volume among all New France posts.

As Colonial European nations fought for control of the continent, the fort fell under British rule, followed by a very brief Spanish occupation. Native groups struggled to hold on to their land. By the end of the 18th century, American control of the area was established. The fort was abandoned but survived in community memory. A commemorative boulder for the post and a granite cross for Father Allouez were erected along the river. In 1998, local group Support the Fort invited Western Michigan University to conduct archaeological investigations. This work continues to reveal information about the fort and its residents.

In 1829, the village of Niles was platted. First called Pawating, it was renamed in honor of Hezekiah Niles, a Baltimore publisher whose weekly news magazine had national influence. Water-powered mills on the Dowagiac and St. Joseph Rivers fueled early industry, while fertile land supported agriculture. Goods were shipped on the river, and a bustling downtown soon filled with shops, inns, taverns, doctors, manufacturers, schools, and churches.

The arrival of the railroad in 1848 spurred further growth. Smaller surrounding villages like Bertrand were left behind. By 1859, Niles was incorporated as a city, complete with police, fire, utilities, and government services. A library, museum, theaters, bands, and civic organizations enriched community life, while the St. Joseph River provided both transportation and recreation. Resorts flourished at nearby Barron Lake. Early Niles citizens worked with city officials to build bridges and pave roads. New schools were built, and quality educators drew students here from across the region.

Niles played a role in the Underground Railroad. Amable LaPierre operated the Eating House at the Michigan Central Railroad Depot. LaPierre was known to provide meals for freedom seekers. Some who had escaped slavery moved on to Canada. Others settled here and established the Second Baptist Church (Mount Calvary today), Franklin African Methodist Episcopal Church, a "colored school" (Ferry Street School), and Michigan's first African American Masonic Lodge.

Historic markers on Ferry Street commemorate these landmarks. Descendants of early African American families still live in Niles today.

Early businesses like the Hunter Ice Company and the Volant Mill defined Niles as an entrepreneurial center. By the 20th century, industries with global reach put Niles on the map. Kawneer transformed storefronts with aluminum framing. Simplicity Pattern Company revolutionized home sewing with affordable patterns. Since 1871, the French Paper Company (formerly Michigan Wood Pulp Company) has produced specialty paper at its riverfront mill, powered by its own hydroelectric dam.

Though industry has since declined, its legacy endures. The Plym family, Kawneer Company founders, purchased and donated the Castle Rest mansion to create Pawating Hospital, gave land for Plym Park, and helped fund a new, larger library. In 1882, the prominent Chapin family built a grand mansion. Purchased by the city in 1933 for just $300, with the condition it be used for civic purposes, the Chapin Mansion served as city hall until 2012 and is now part of the Niles History Center.

Notable figures with ties to Niles include John and Horace Dodge, founders of the Dodge automobile company; Aaron Montgomery Ward, pioneer of mail-order retail; and sportswriter and humorist Ring Lardner. Artist and activist Lottie Wilson became the first African American to attend the School of the Art Institute in Chicago. Her painting, "President Lincoln with a Former Slave," depicts the historical meeting of Lincoln and civil rights advocate Sojourner Truth, and was displayed by President Theodore Roosevelt at the White House.

In choosing the photographs for this book, the authors looked through the archives of the Niles History Center, with a few images from other institutions where it was deemed necessary to tell a story. With a limit on the final number of photographs that could be included, images were chosen for their clarity, interpretive quality, and place in Niles history. A choice was made to limit images to before the mid-20th century, with a few exceptions. The *Images of America* series prefers historical photographs. Later stories will have to wait for another time: 20th-century Main Street development, the loss of industry, the Eleventh Street corridor, the Vietnam and Gulf Wars. Also for another day are Niles's modern-day celebrities like rock star Tommy James, who topped the charts with hits like "Hanky Panky" and "Crimson and Clover."

Even as Niles evolved, it held tightly to its history. Much of the downtown retains its original character, with landmarks like the Four Flags Hotel, Carnegie Library, and 1910 Post Office still standing. The grand 1892 railroad depot continues to welcome travelers, while monuments and cemeteries honor veterans and significant places. Long-standing events like the Apple Festival and Hunter Ice Festival have deep roots in the community. The Pokagon Band of Potawatomi has reaffirmed its sovereignty as a federally recognized tribe. This volume does not attempt to tell the entire story of Niles, but rather to share snapshots of its past—captured in the present—for future generations.

One

From Fort to Village

Before Europeans arrived, this land was known as Pawating, meaning "place of the rapids." Native people traveled along the river now called the St. Joseph and its tributaries. They also moved across an east–west trail, worn flat by roaming herds of mammoths, mastodons, and later buffalo.

The French government established Fort St. Joseph on the eastern bank of the river in 1691, at the site of an earlier Jesuit mission. The fort served as a mission, garrison, and trading post. After the Seven Years' War (1756–1763), it fell under British control. Within a year, Native groups allied under the Odawa leader Pontiac rose in rebellion, attacking several forts, including Fort St. Joseph. The English commander, Ensign Francis Schlosser, was taken prisoner. Although the British briefly used the fort during the American Revolution to arm allies fighting American colonists, they never permanently regarrisoned it.

In 1781, a French and Native American expedition from Spanish-controlled St. Louis claimed Fort St. Joseph for the King of Spain. They plundered the site and raised a Spanish flag, but remained only a single day. This incident earned Niles the nickname "City of Four Flags," referencing French, British, Spanish, and American rule. The title, however, overlooks both Pontiac's influence and the earlier sovereignty of Native nations.

During the early 1800s, trade continued with the establishment of Bertrand, just south of present-day Niles. In the 1820s, Isaac and Christiana McCoy founded a Baptist mission and school west of the river. Their school for the Potawatomi taught agriculture and trades, often at odds with traditional lifeways. The McCoys left in 1829, advocating for Potawatomi removal, and the mission closed the following year. At a time when many Native peoples were being forced from their homelands, Chief Leopold Pokagon successfully negotiated for his people to remain. The Pokagon Band of Potawatomi, whose presence here stretches back centuries, received federal reaffirmation in 1994.

Although settlers first referred to the village as Pawating, in 1829 it was renamed Niles, in honor of Hezekiah Niles, a well-known Baltimore newspaper editor. Early industries developed as settlers constructed dams to harness waterpower for their mills. As more people moved westward, a thriving town developed along the St. Joseph River.

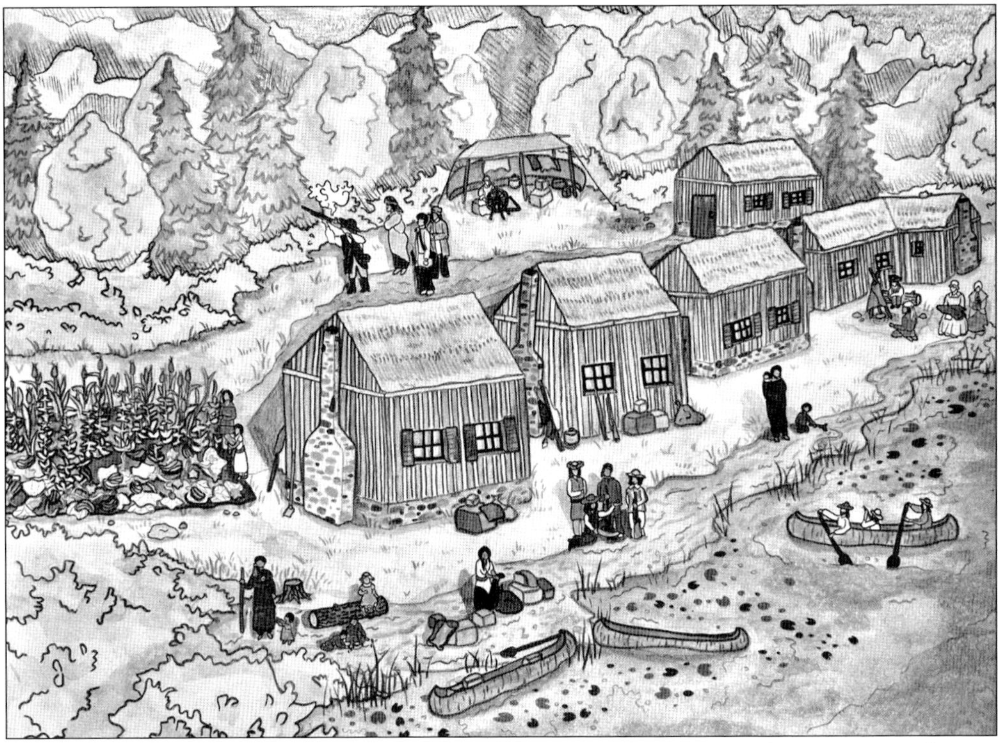

Fort St. Joseph was a mission, garrison, and trading post built by the French in 1691. No plans or maps are known to exist. This drawing, based on archaeological investigations conducted since 1998, imagines what the fort might have looked like. Traders arrive with goods in canoes, greeted by fort residents. Evidence of houses, gardens, and fireplaces has been discovered. (Drawing by Paige Pemble, courtesy of the Fort St. Joseph Archaeological Project.)

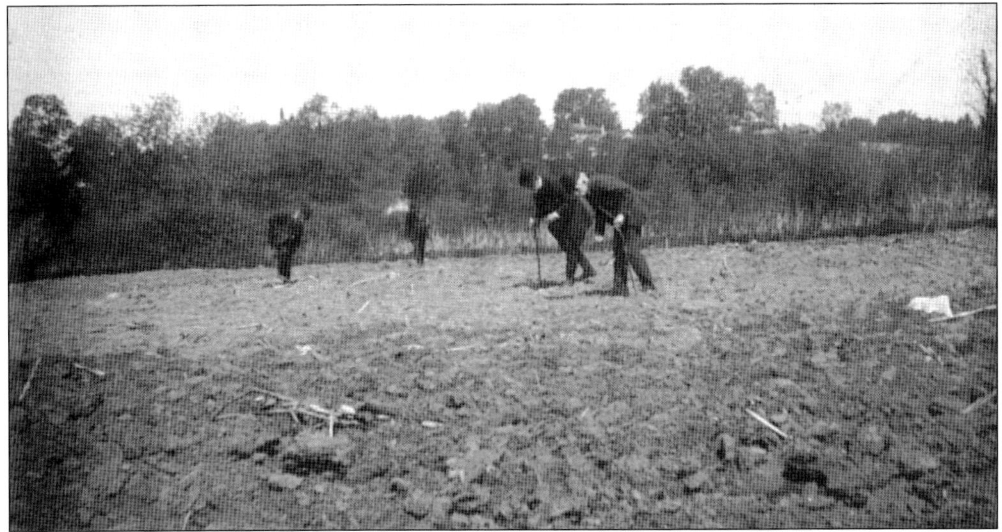

In the early 20th century, a group of avocational archaeologists scoured the area of the old fort looking for artifacts. Called the Miami Cross Society, the group consisted of Messrs. Lombard, Beeson, Smith, and Crane. They operated under the motto "us four and no more." Many of the society's finds ended up in the collections of the Fort St. Joseph Museum.

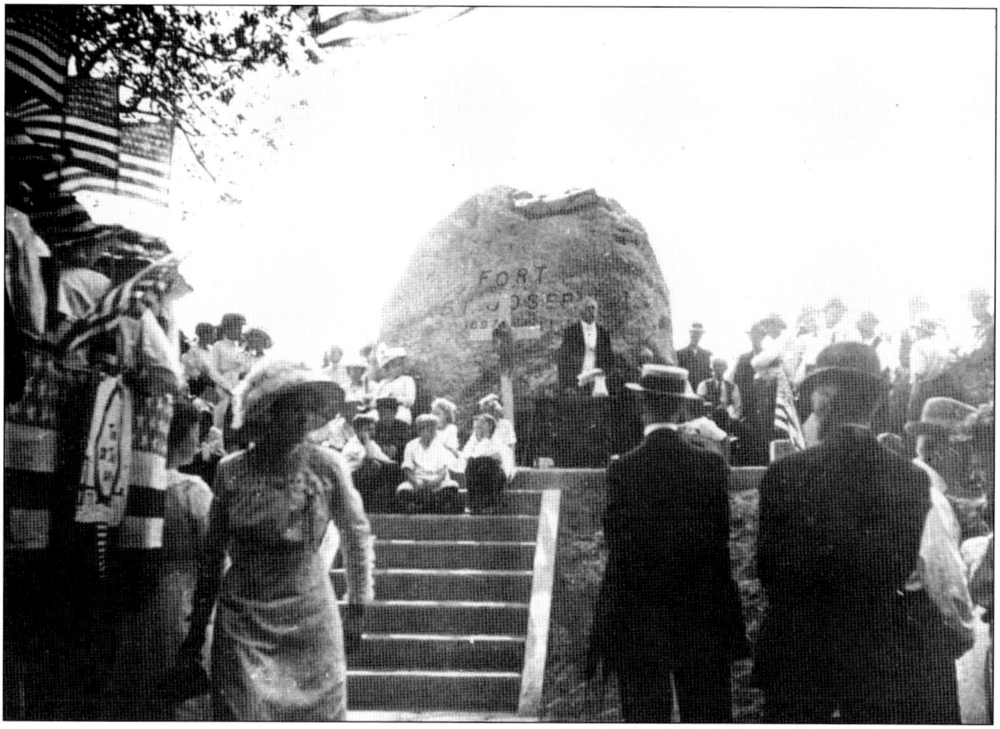

On July 4, 1913, a ceremony with much pomp and circumstance dedicated a large boulder placed to mark the location of Fort St. Joseph. The 12-foot-tall boulder was sourced from a local swamp. It took more than two years of planning to move and install, with a total cost of $1,000. A parade through town, costumed reenactors, and speeches drew the people of Niles to the momentous event.

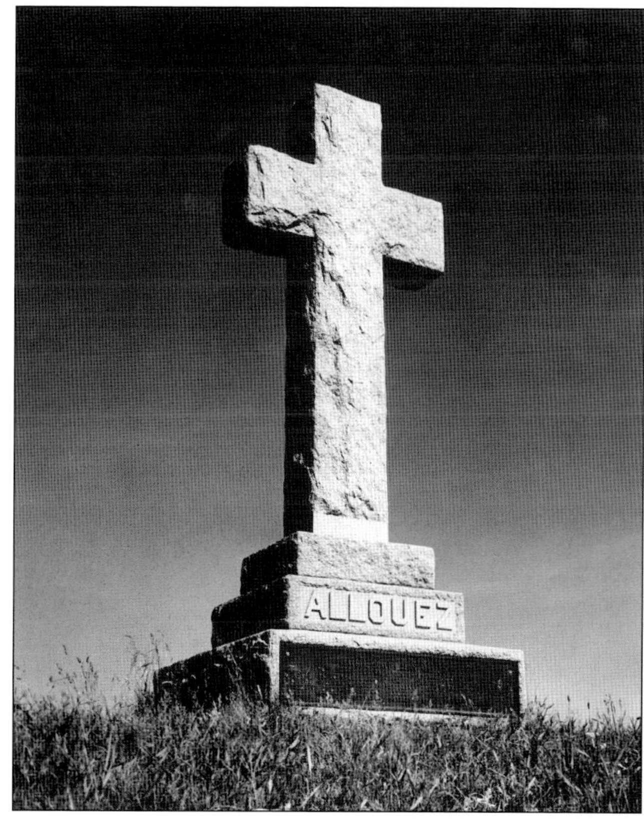

Father Jean Claude Allouez was a Jesuit priest who helped establish a mission here in the 1680s with the intent of bringing Catholicism to Native people. He died in 1689 at the mission. A wooden cross marked his grave site. In 1918, the Women's Progressive League placed a more permanent granite cross on a hill near Fort and Bond Streets.

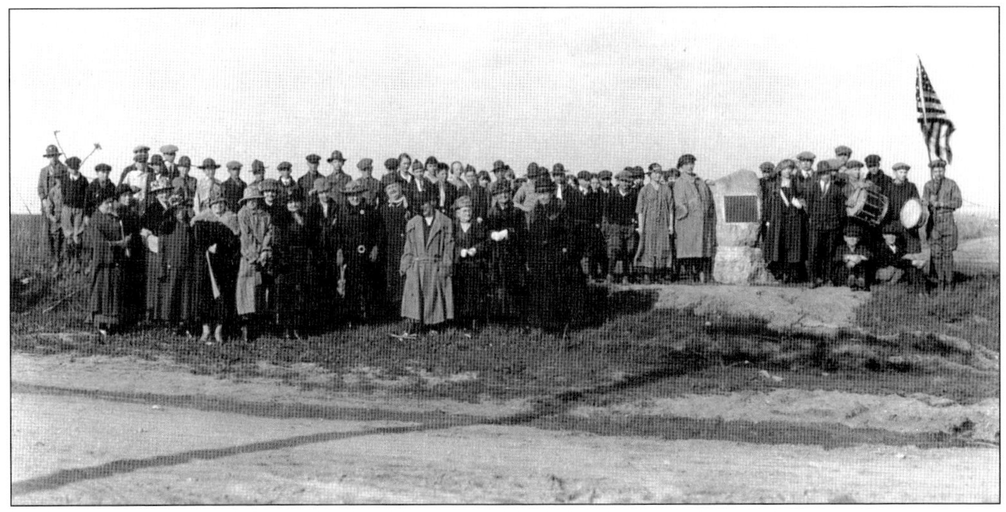

Isaac and Christiana McCoy started the Carey Mission and its school in 1823. A plaque marking the location of the Carey Mission stands at Niles-Buchanan and Philip Roads. The Daughters of the American Revolution's Fort St. Joseph Chapter, with help from local Boy Scouts and Campfire Girls, dedicated the plaque in 1923 on the centennial of the mission's founding.

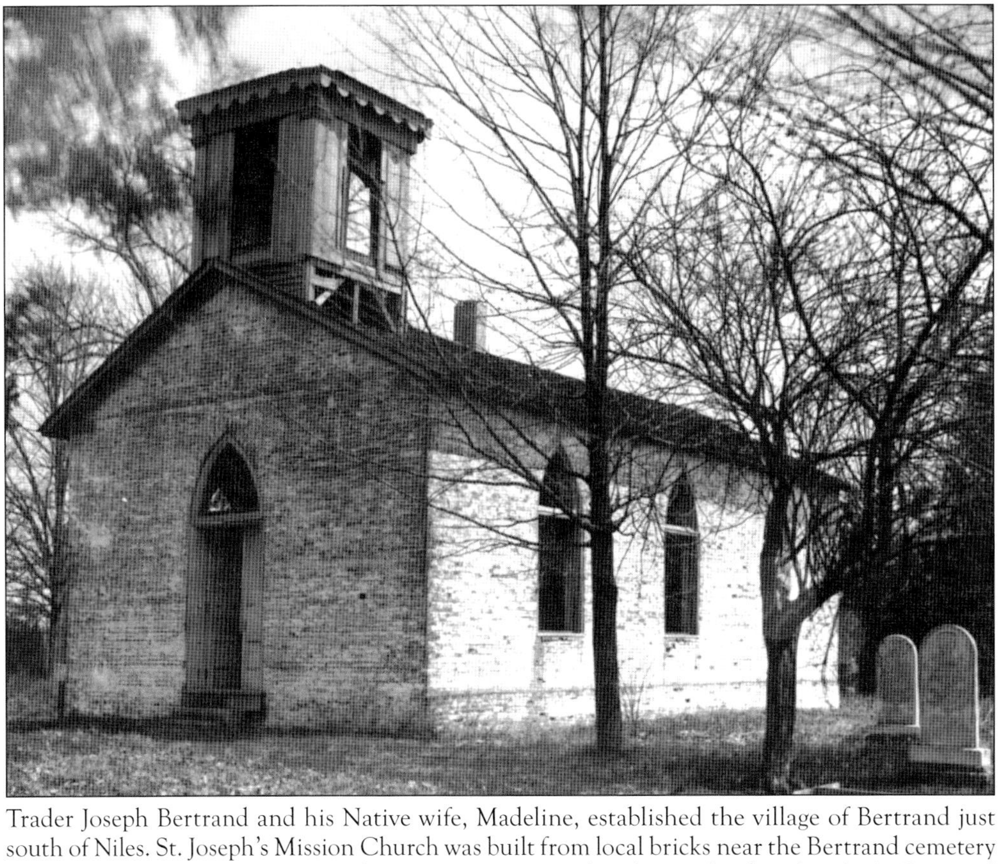

Trader Joseph Bertrand and his Native wife, Madeline, established the village of Bertrand just south of Niles. St. Joseph's Mission Church was built from local bricks near the Bertrand cemetery on a bluff overlooking the St. Joseph River in 1837. It replaced an earlier log church. As Bertrand declined, the church fell into disrepair and was razed in 1911.

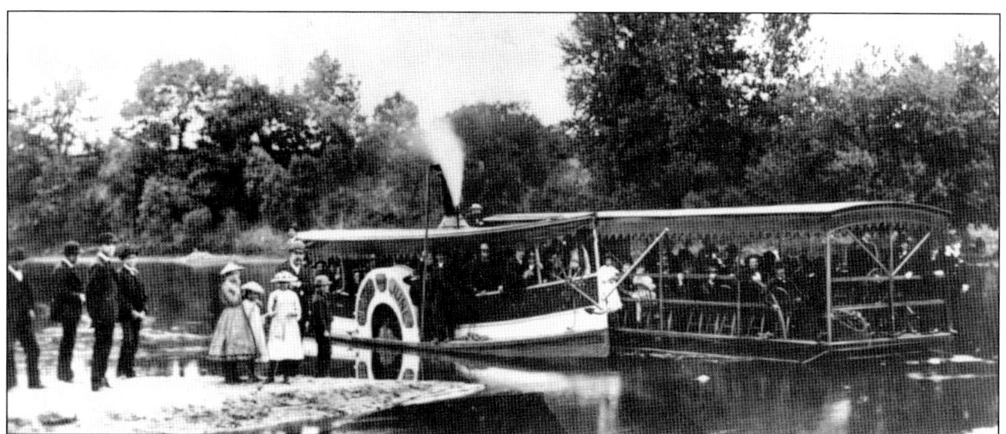

One of the last steamships to ply the St. Joseph River, the *Nettie-June* was a sidewheel steamer owned by Jennie and Andrew Carothers of Buchanan. Built in 1882 and named for their two daughters, the *Nettie-June* made trips on the river between Niles and St. Joseph, carrying passengers for a fare. The steamer with tow, shown here, could transport 100 passengers.

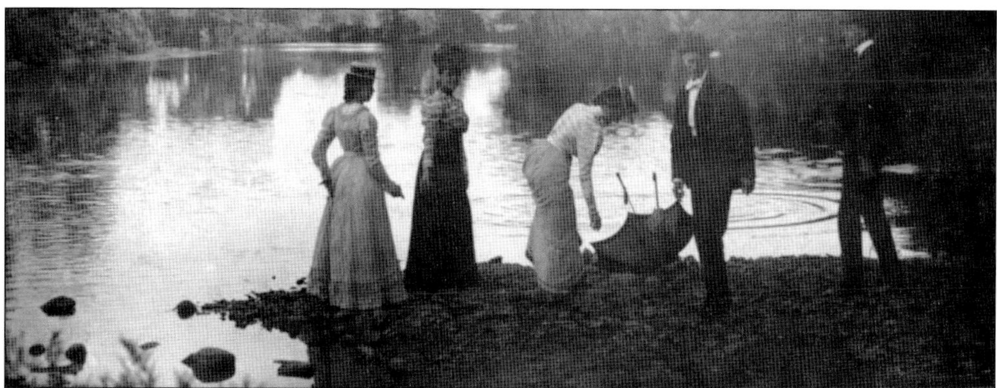

Niles has benefited not only from the St. Joseph River's waterpower but also from the recreation and scenery it provides. Here, early Nilesites, from left to right, Belle Root, Pet Reddick, Ruth Reddick, Walter Smith, and Fred Dean enjoy a day on the riverbank. The photograph dates to the 1890s.

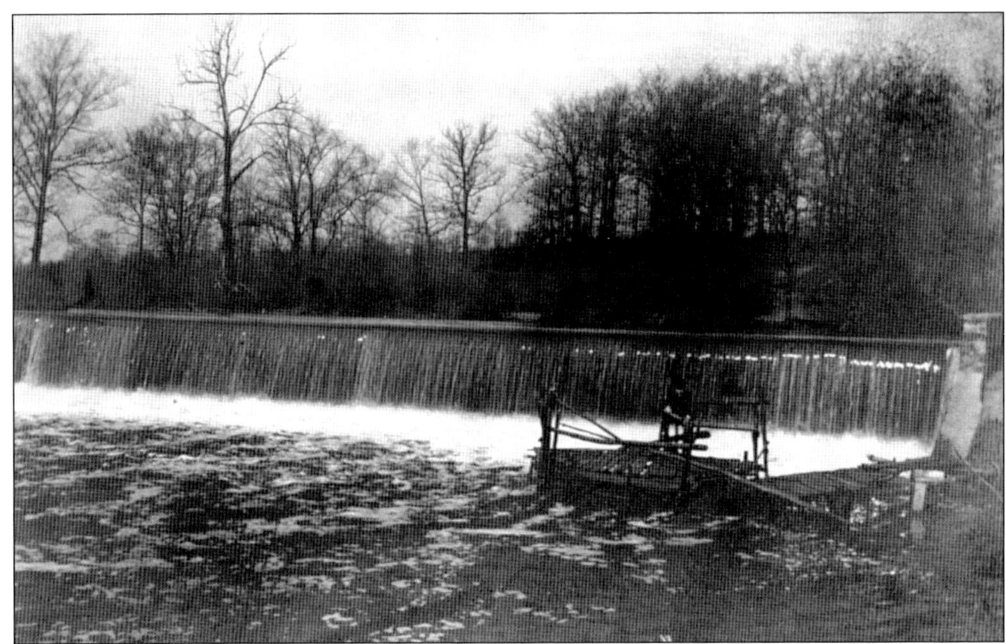

Early efforts to dam the St. Joseph River proved difficult. Settlers instead focused on the Dowagiac River, a narrow tributary just to the north. Here is a photograph of one such dam, perhaps the Lacey Dam, which powered early mills. The location is identified as today's Plym Park at the bottom of Hill Street. Obed P. Lacey was one of Niles's earliest settlers, arriving in 1828.

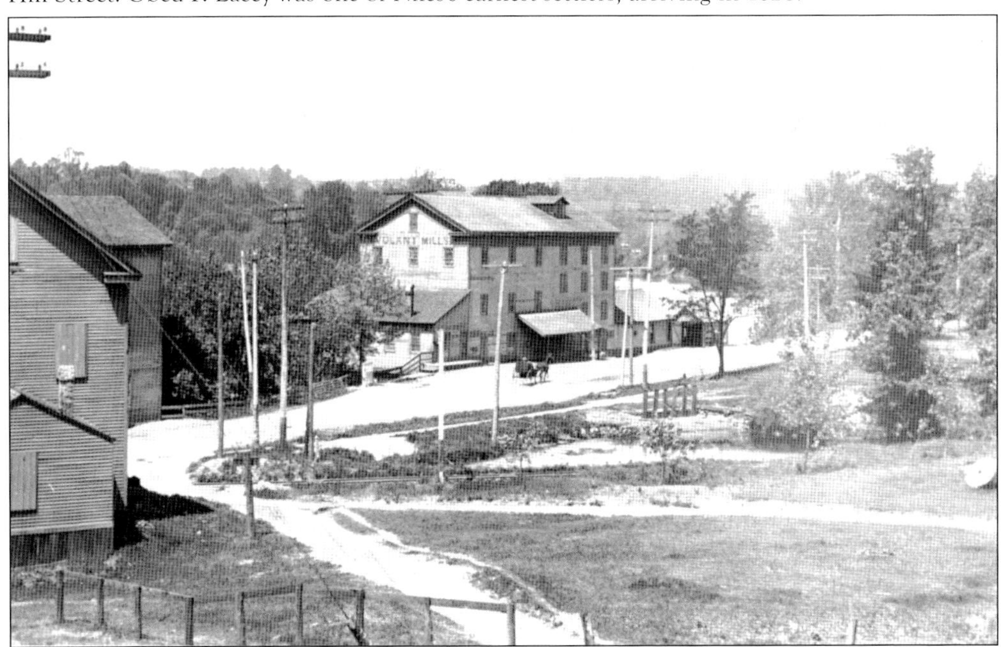

Samuel Finley built Volant Mill in 1847. Volant was a large flour mill on the east bank of the St. Joseph River, just north of downtown. A raceway from the Dowagiac River powered the mill. In 1881, ownership of Volant and several smaller mills was consolidated under the Niles Milling Company. Small local mills declined during the early 20th century with the rise of larger production mills and imported wheat.

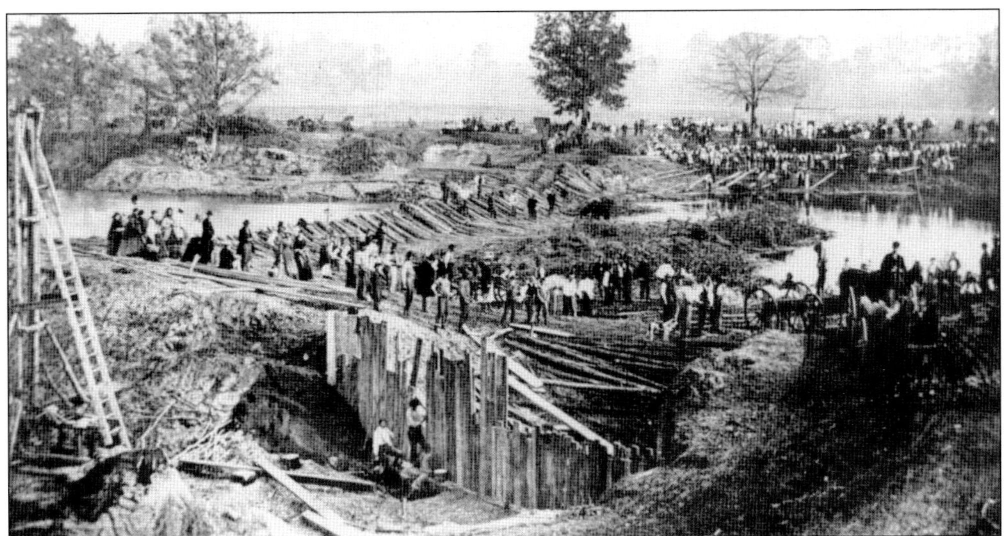

Dam construction on the St. Joseph River was a major undertaking. After several challenges, in 1871, the Niles Manufacturing Company completed the raceway, headgates, and other work needed for Niles's first permanent dam on the river, just south of the bustling downtown. Several businesses leased rights to this newly harnessed waterpower.

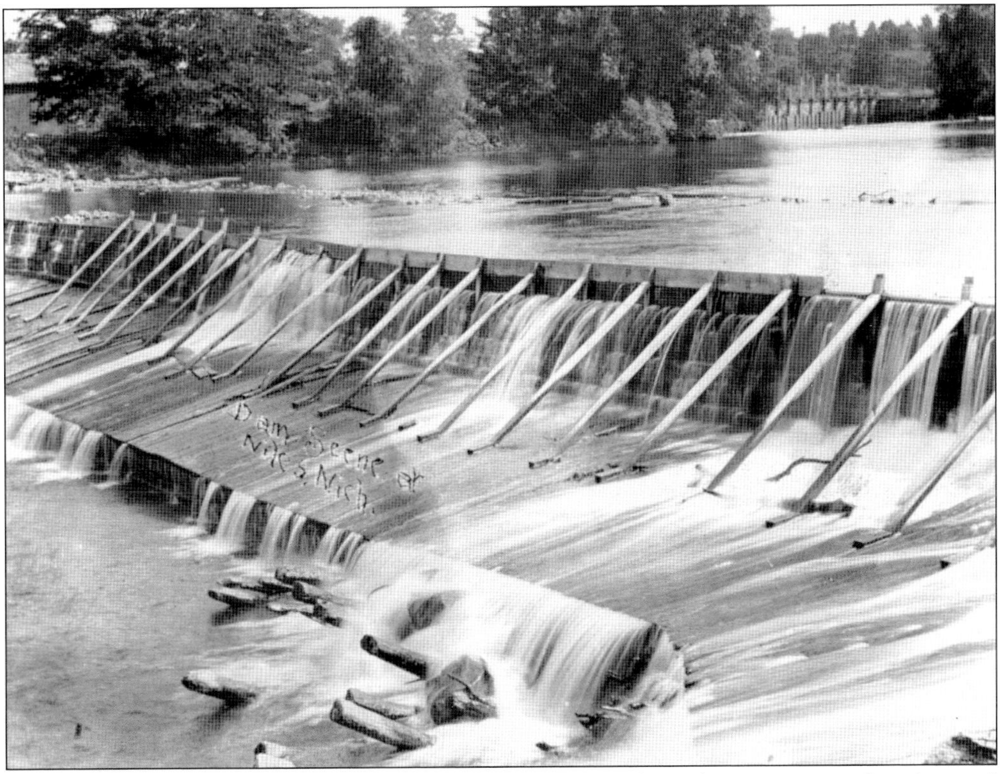

W.A. Bolton took this photograph of the St. Joseph River dam, showing its wood timber and earth construction. The headgate of the mill race on the east side of the river is visible in the background. The mill race was a channel constructed to power mills along the river. A concrete structure replaced this dam by the end of the 19th century.

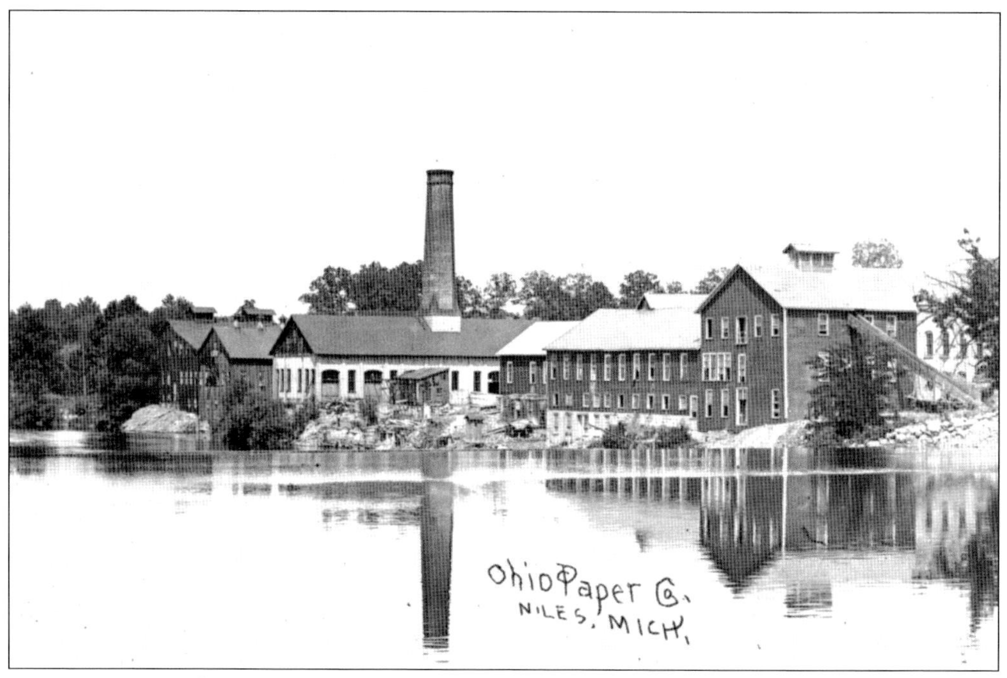

Benjamin Faurot from Lima, Ohio, started the Ohio Paper Company in the early 1880s. The company operated on the east bank of the St. Joseph River, next to the Niles Paper Company. Ohio Paper produced a patented material used in paper boxes and was one of the few companies in the country to manufacture pulp-lined strawboard.

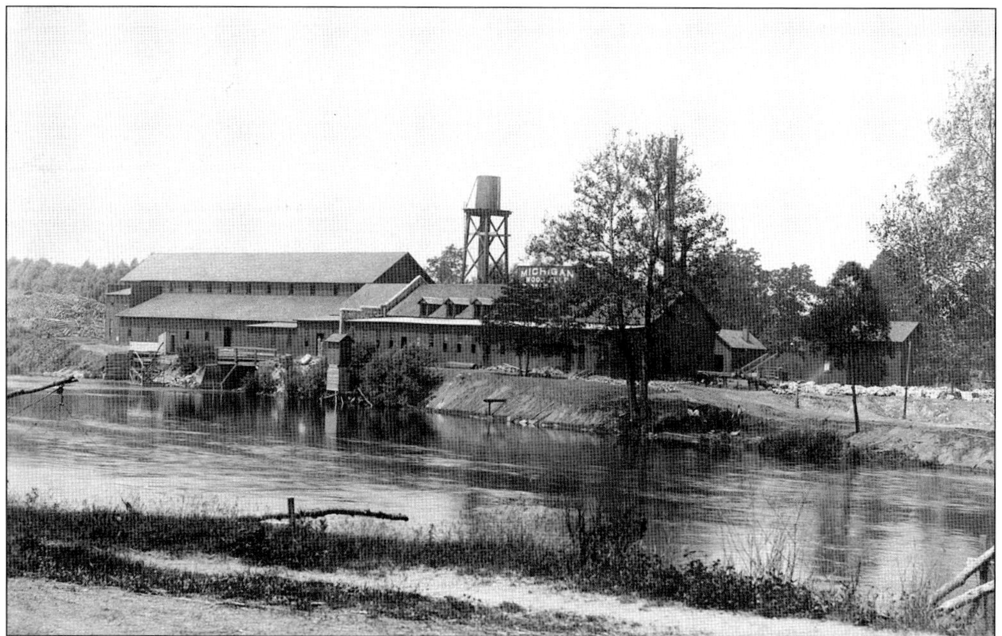

In 1871, Joseph W. French and J.B. Millard organized the Michigan Wood Pulp Company, utilizing power from the newly constructed dam on the St. Joseph River. Pictured here in 1891, the factory sat on the west side of the river. A sign reading "Michigan Wood Pulp Company" is just visible behind the tree on the building's right side.

In 1905, Michigan Wood Pulp became the French Paper Company under the sole ownership of the French family. French Paper Company joined Finch Paper Holdings of New York in 2019 and continued operations in the same location, making specialty paper products. A hydroelectric dam still powers the plant. This photograph dates to shortly after 1920.

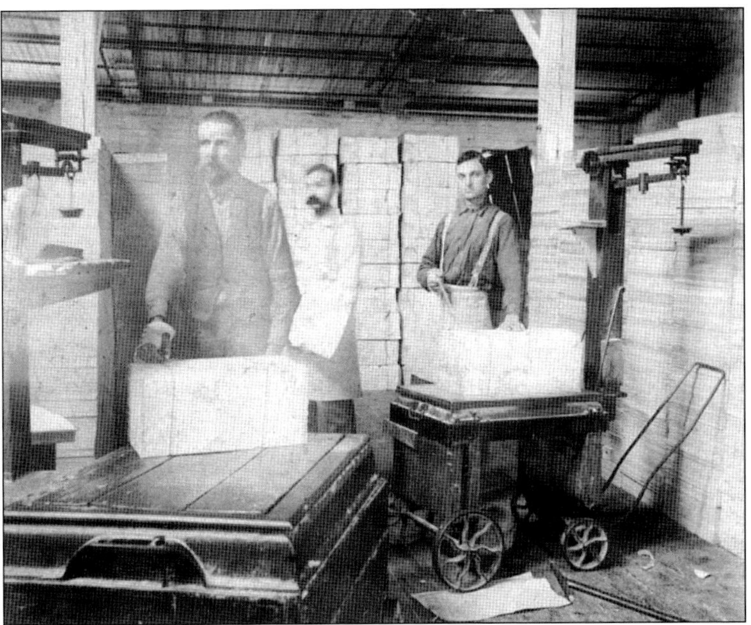

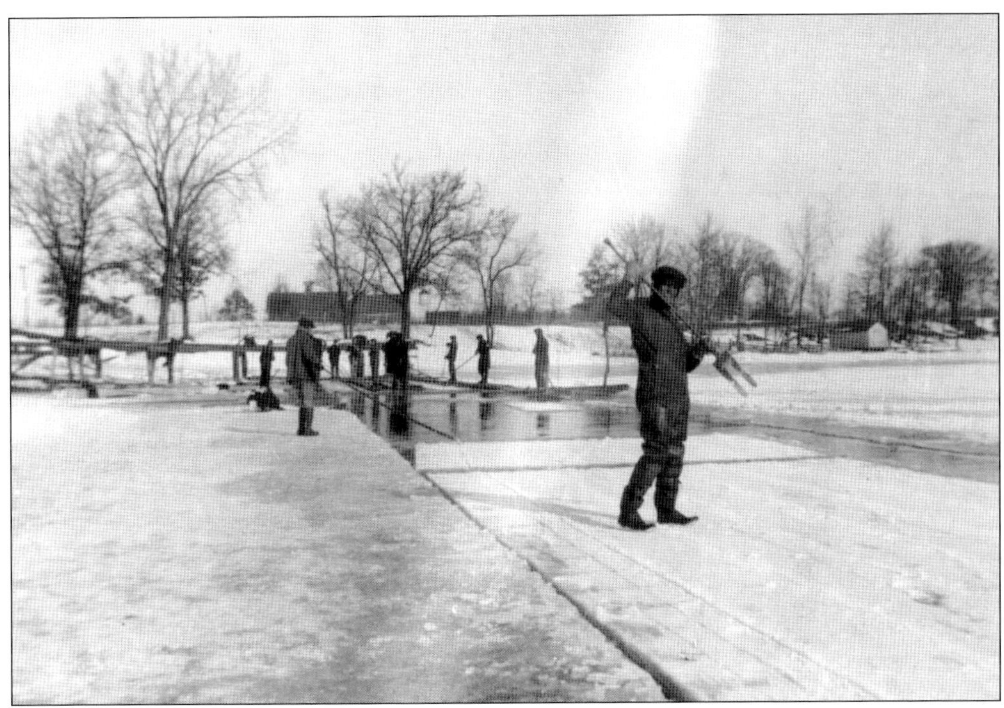

Before artificial ice production, ice harvesting was a major industry. Ice needed to be at least nine inches thick before it was cut and sent downstream to an ice house. Ice blocks were covered with hay to prevent melting. Properly insulated blocks could last a whole year! The Michigan Central Railroad delivered ice to Niles customers and to every stop between Michigan City and Kalamazoo. (Courtesy of the Dowagiac Area History Museum.)

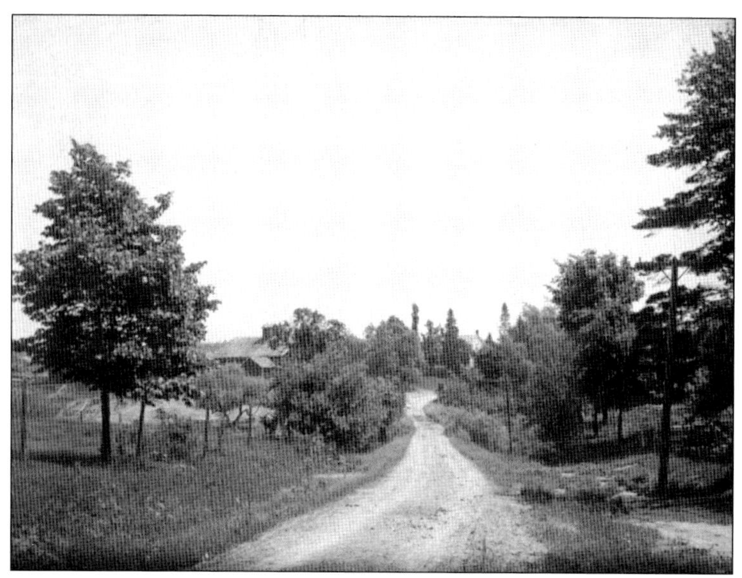

Agriculture flourished around Niles's urban center. Robert Alexander Walton, a Civil War veteran, bought farmland a few miles north of town in the 1860s. Robert married Melvina Ribble, and they developed around 200 acres of land. When Robert and Melvina retired and moved to town, their son Herbert took over the farm. Present-day Walton Road runs through this area.

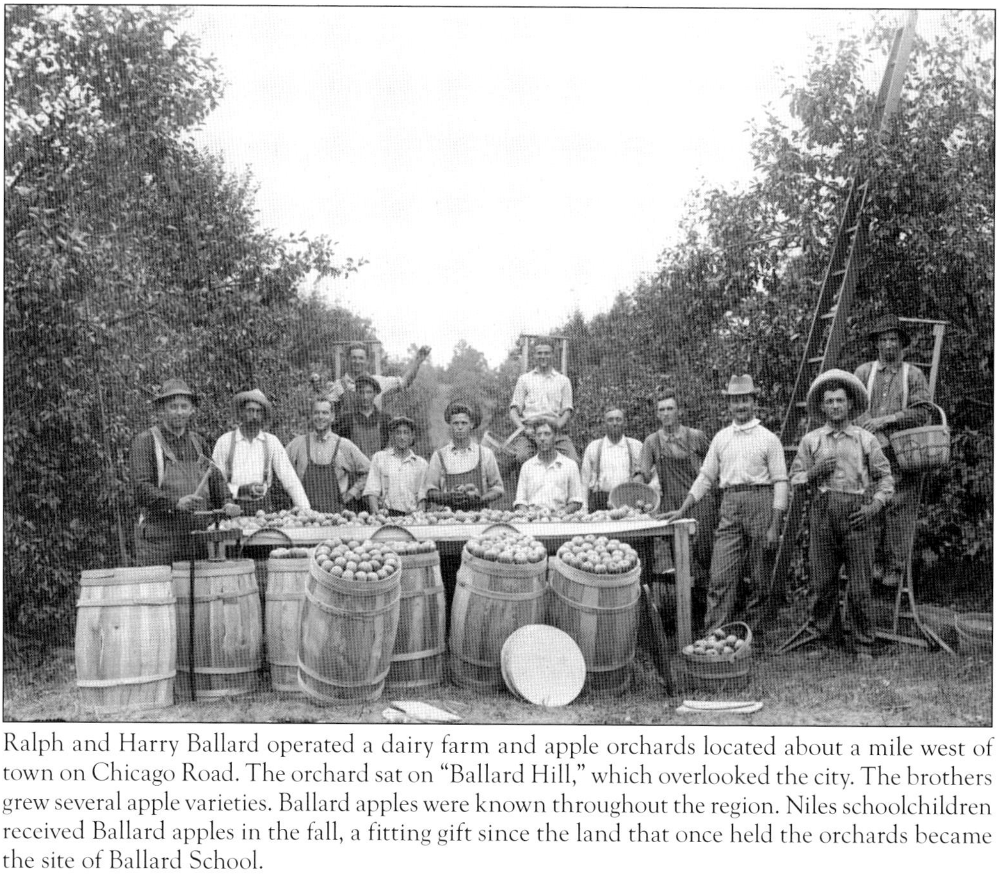

Ralph and Harry Ballard operated a dairy farm and apple orchards located about a mile west of town on Chicago Road. The orchard sat on "Ballard Hill," which overlooked the city. The brothers grew several apple varieties. Ballard apples were known throughout the region. Niles schoolchildren received Ballard apples in the fall, a fitting gift since the land that once held the orchards became the site of Ballard School.

Ralph Ballard created his *Map of Indian Trails* using government surveys, archaeological maps, and local lore. The map was compiled for the Fort St. Joseph Historical Association in 1939. Before settlement, the area served as a crossroads for native groups who used land trails and also traveled on the river and its tributaries. Ralph's wife, Mary, was also a historian and teacher who helped with research and programs.

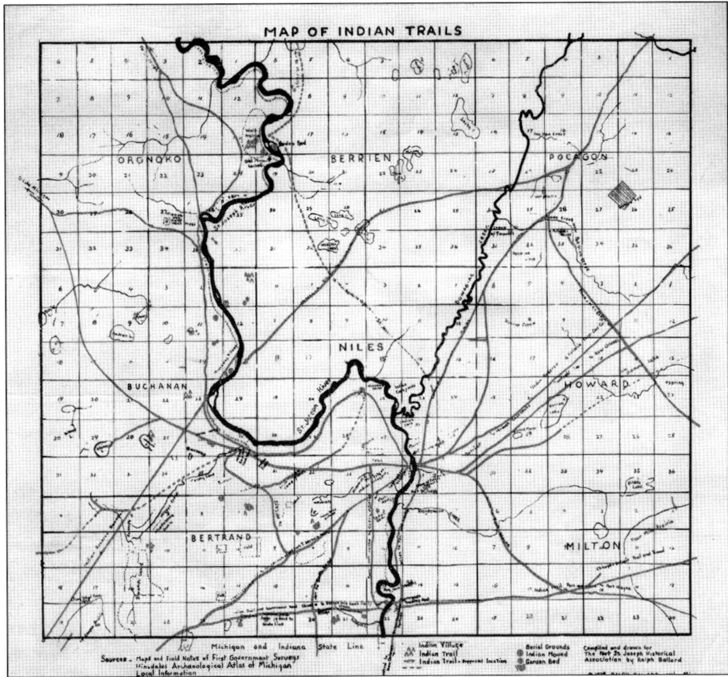

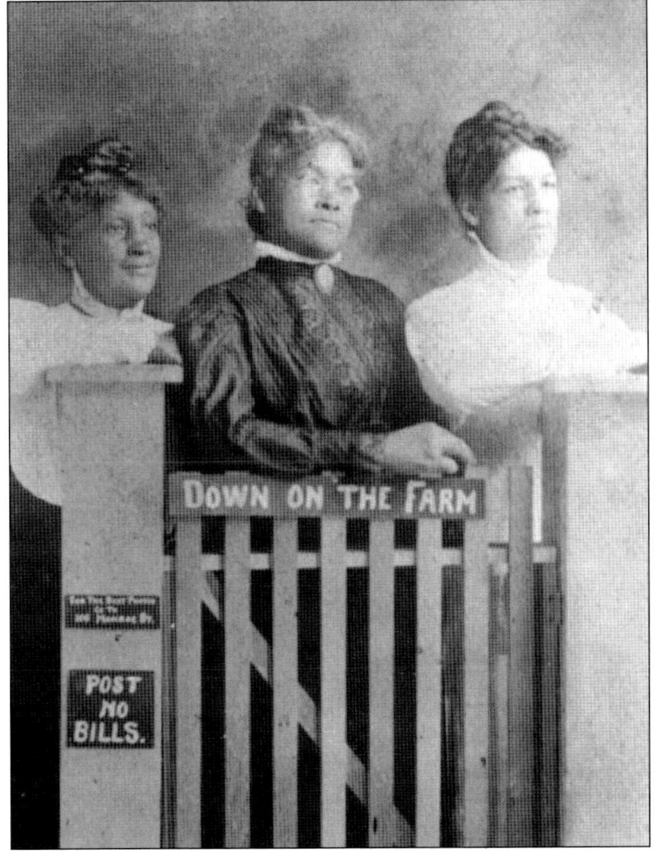

Pasquel Finley moved his family to Niles from Ohio in the 1840s, believing the area safer for African Americans. He worked for the Michigan Central Railroad, grading land for the new rail lines, and later purchased land near the present-day airport and became a cattle farmer. Pictured here from left to right are Pasquel's daughters: Ophelia, Permelia, and Mary Finley. Finley descendants still live in Niles.

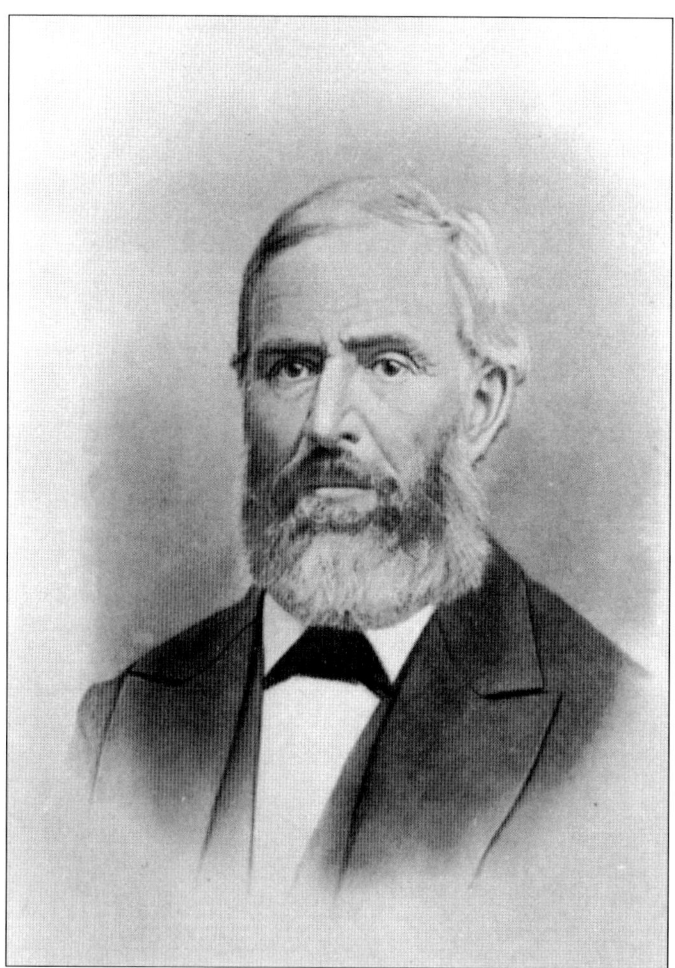

Jacob K. Brown moved to Niles in 1831 at the age of 21. He served as a Village of Niles trustee and was a builder by trade. According to his obituary, he oversaw the construction of 21 schoolhouses, a dozen mills and at least 50 stores, dwelling houses, and other buildings. The family of Jacob K. Brown still calls Niles home.

In 1867, Henry Pike opened the Pike House at the southwest corner of Main and Fourth Streets. Built by local contractor Jacob K. Brown, the Pike House featured modern amenities such as electricity and private bathrooms. Carriages transported patrons arriving by railroad from the depot to the hotel. The Pike House was demolished in the 1920s to make way for the Four Flags Hotel.

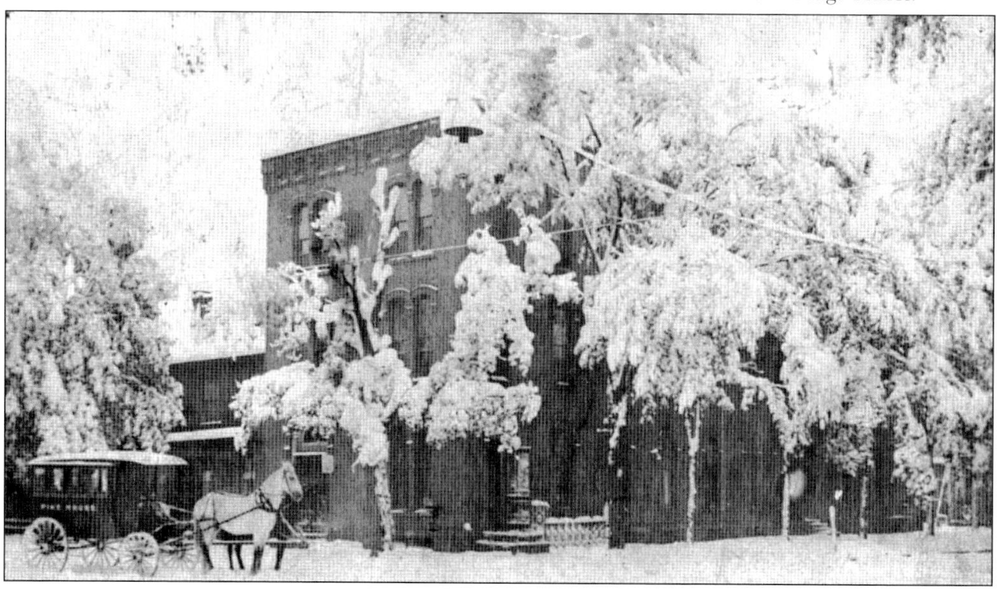

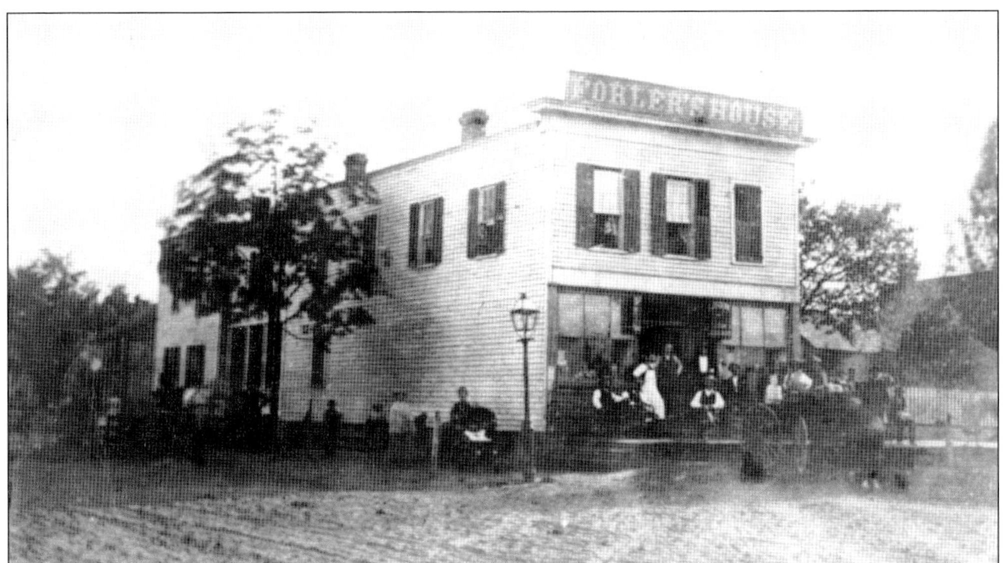

George Forler opened a boardinghouse and grocery at Fifth and Wayne Streets in the mid-1860s to accommodate visitors traveling to Niles on the railroad. The structure shown here burned in 1889, and a new brick building replaced it. After Forler's retirement in 1895, the Forler House, later called various names including the Powell Hotel and the Borden House, continued under different owners until its demolition in 2000.

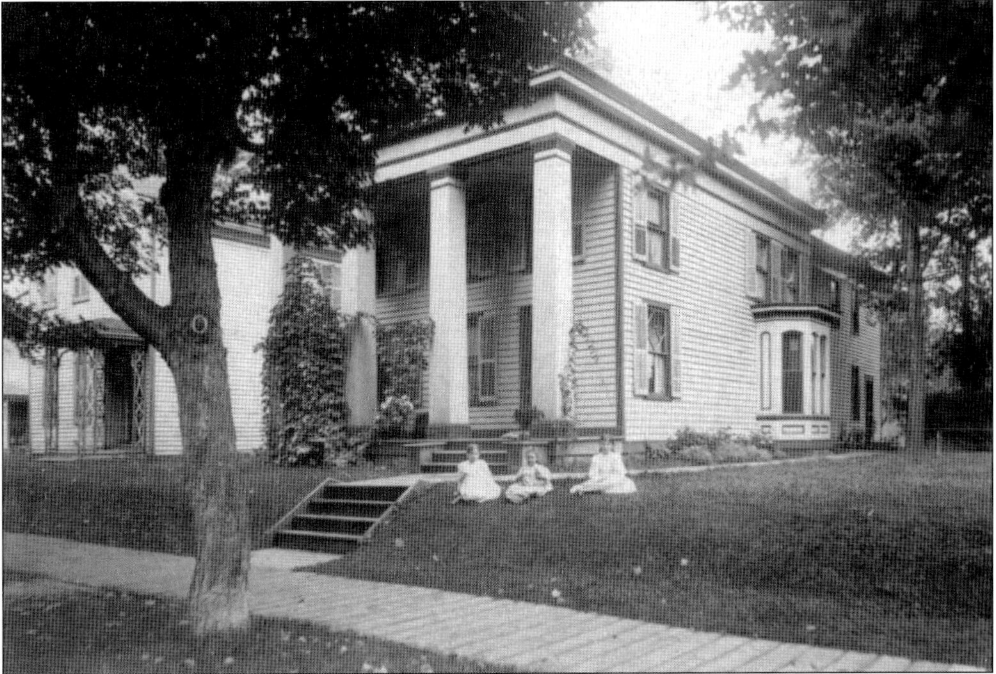

Here is the home of Edwin Clarence Griffin, a prominent Niles businessman and the son of Capt. Eli Griffin, who was killed during the Civil War. The home was located on North Fourth Street. This 1890s photograph identifies the three girls as Jeanne Griffin, Harriet Clare Griffin (E.C. Griffin's daughters), and Viola Humiston. Jeanne Griffin became a librarian and served as delegate-at-large for the Michigan Division of the United Nations Association.

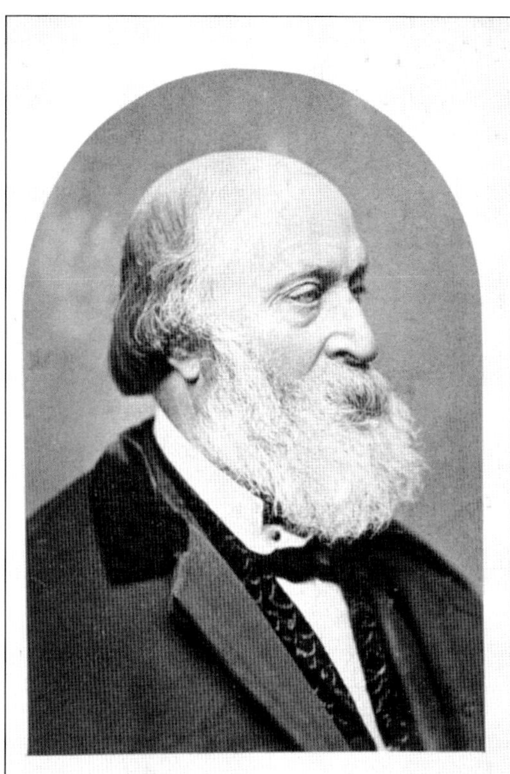

George Jerome, a charismatic local naturalist, built a large estate on South Third Street in the early 1870s. Known as Sabine Farms, the property featured a grand home for the family and sprawling grounds with gardens, ponds, and a private fish hatchery. In 1873, the state appointed Jerome as the first commissioner of state fisheries, a precursor to the modern Michigan Department of Natural Resources.

This photograph shows a 1929 Pokagon Band Tribal Council meeting and election that took place near Bertrand. Europeans first encountered the Neshnabek (Potawatomi) people living here in the early 17th century. Though many were forced to leave, Leopold Pokagon successfully negotiated for his people to remain on their homeland. His descendants have held on to their language and cultural traditions. The band's sovereignty was reaffirmed in 1994 in a signing ceremony at the White House with Pres. Bill Clinton. (Courtesy of the Pokagon Band of Potawatomi Indians.)

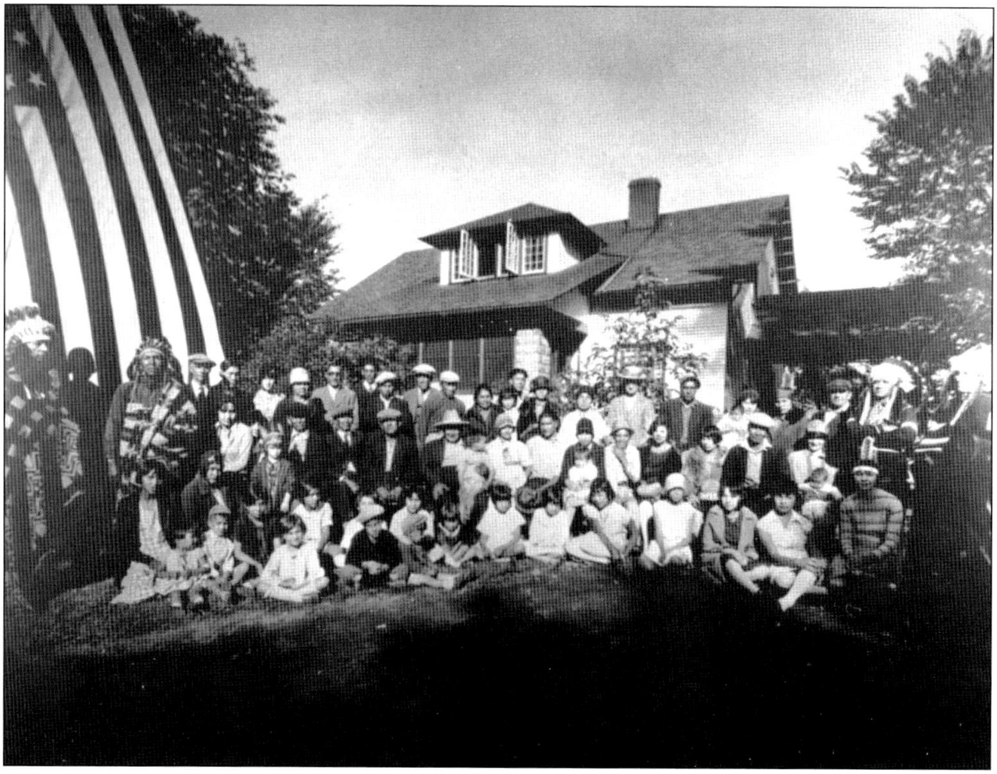

Two

Arrival of the Railroad

Railroads played a major role in America's 19th-century industrial revolution and transportation boom. Intensive political lobbying brought the railroad to Niles in 1848. The Michigan Central Railroad built a line to the Chicago gateway through southern Berrien County, instead of the originally planned route from St. Joseph. Niles rejoiced and hailed the arrival of the first train.

A second line, the Airline Railroad, opened in 1871, running from South Bend through Niles via Barron Lake and on to Jackson, Michigan. The "Big Four" Railroad followed in 1882. Short for Cincinnati, Chicago, Cleveland, & St. Louis (also abbreviated as CCC&StL), it ran along the river and continued north to Benton Harbor. In 1919, Michigan Central moved its classification yards from Michigan City to Niles, making the city a major freight division center. The new yards and engine terminal created more railroad jobs and further defined Niles as a railroad city.

For local travel, the South Bend & Southern Michigan Railway built an electric rail line. Known as the "interurban," it connected Niles and South Bend with other towns across southwest Michigan. After a slowdown during the Depression, railroads came roaring back during World War II. Freight and passenger traffic surged to unprecedented levels, with troops departing Niles by train and the yards working at full capacity. After the war, as automobiles and air travel expanded, railroads struggled.

The Michigan Central, which had operated as a subsidiary of the New York Central Railroad (NYCRR) since the 19th century, was fully absorbed by its parent company in 1930, along with the Big Four. In 1958, the NYCRR opened the computerized Robert Young Yard in Elkhart, Indiana, marking the decline of major freight operations in Niles. However, railroads had already peaked. Despite consolidation efforts, the NYCRR declared bankruptcy in 1970.

Rail travel saw a revival in the 1970s with the creation of Amtrak, the federally funded passenger rail service. Amtrak operates from the majestic depot still standing proudly on 5th Street. Meanwhile, many abandoned rail corridors found new life as recreational spaces through the Rails-to-Trails program.

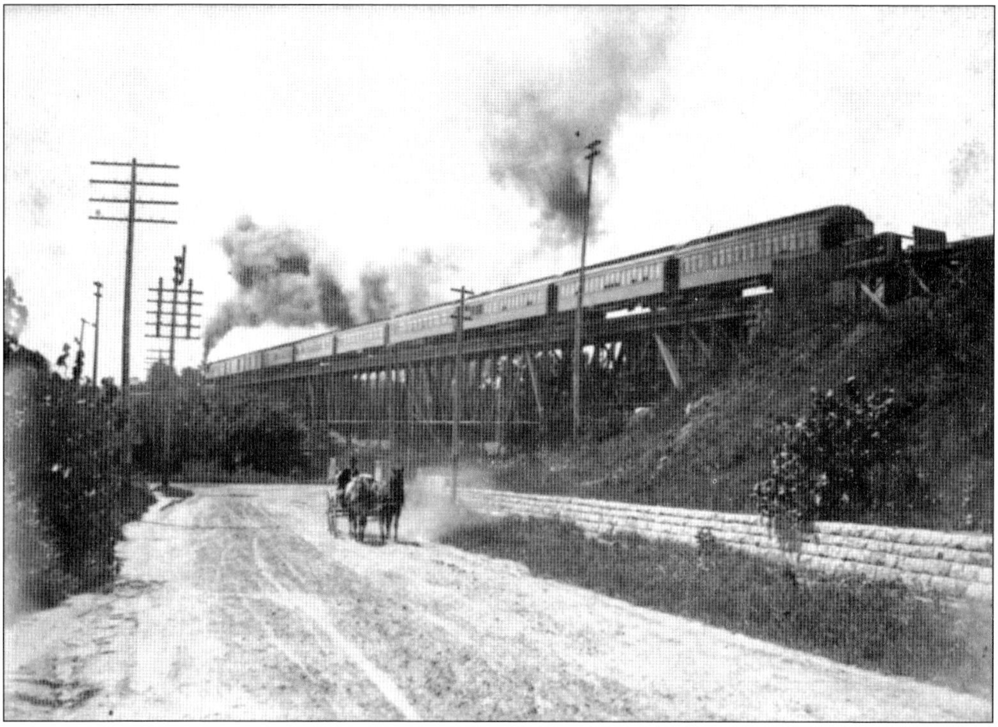

A westbound Michigan Central Railroad (MCRR) train roars along the bridge over the St. Joseph River in about 1890. Below, a horse and buggy travel east along Wayne Street (formerly High Street), perhaps transporting passengers between the MCRR depot and the Big Four depot. One of the utility poles in the background has a semaphore signal for directing trains.

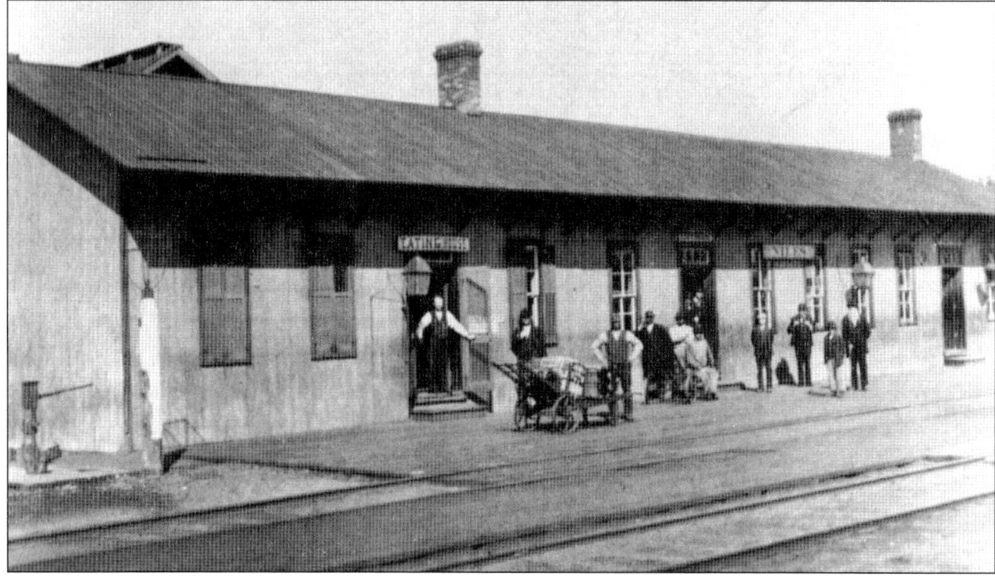

No photographs are known to exist of Niles's first MCRR depot, which was destroyed by fire. A second depot was built in 1873, complete with an Eating House, on the south side of Dey Street. Though planned as a temporary structure, this depot served Niles for nearly 20 years until the present depot was built in 1892.

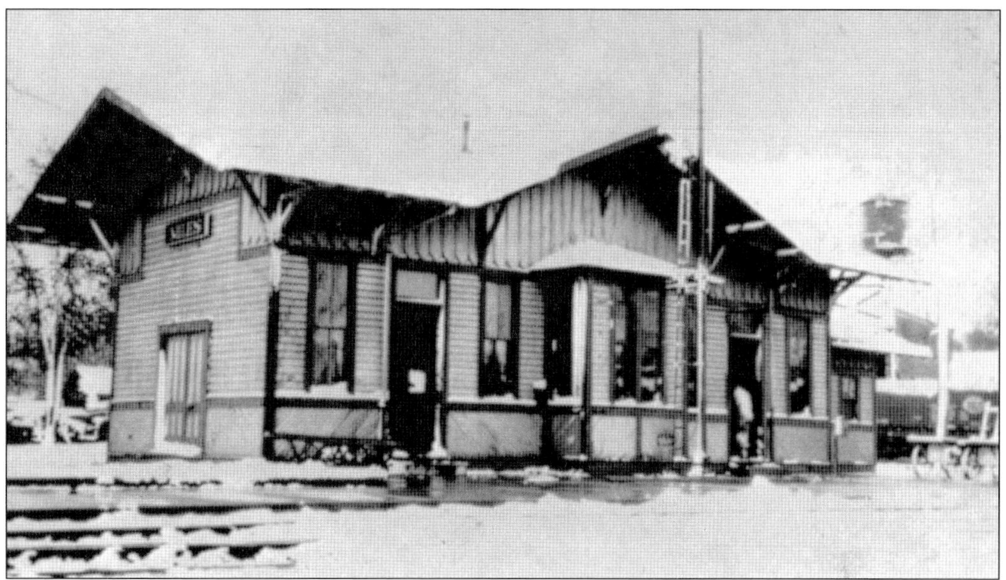

In 1882, the Elkhart, Niles & Lake Michigan Railroad built a line from Niles to Granger, Indiana. The line was extended north to Benton Harbor and then bought out by the Big Four (Cleveland, Cincinnati, Chicago & St. Louis Railroad) in 1890. Passenger service ended in 1933, but freight service continued until 1982. The Big Four Railroad depot was located in today's Riverfront Park.

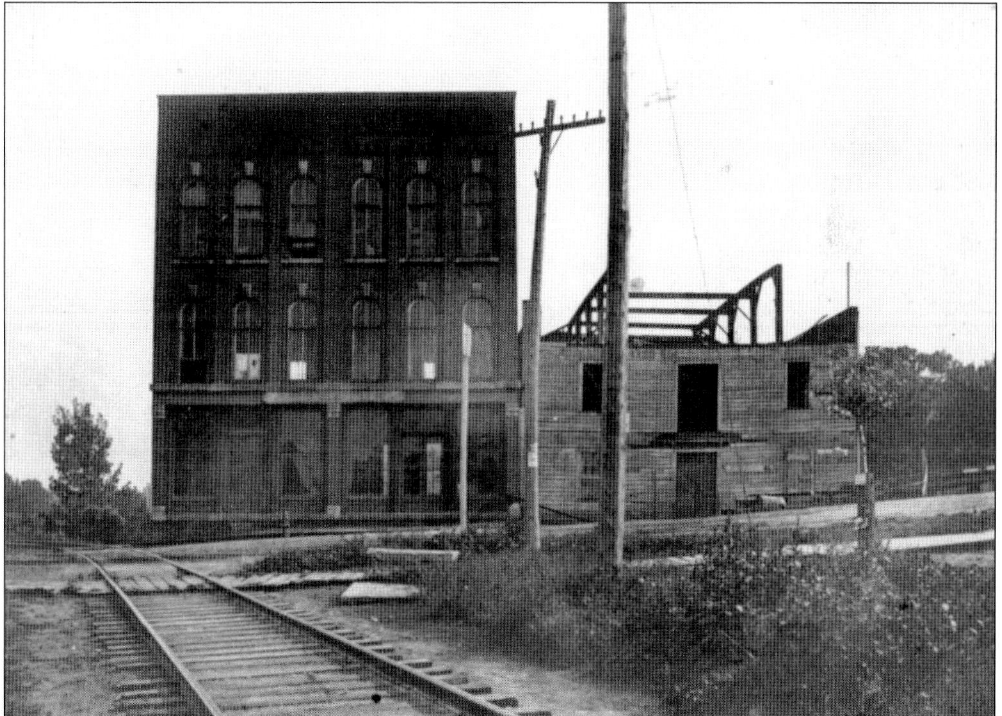

This image shows the Big Four crossing on Main Street looking south. The brick building held the J.S. Tuttle Leather Company and was later used as the city hall. The wooden building was Tuttle's warehouse that stored goods for shipping on the St. Joseph River, which was just to the right of the photograph.

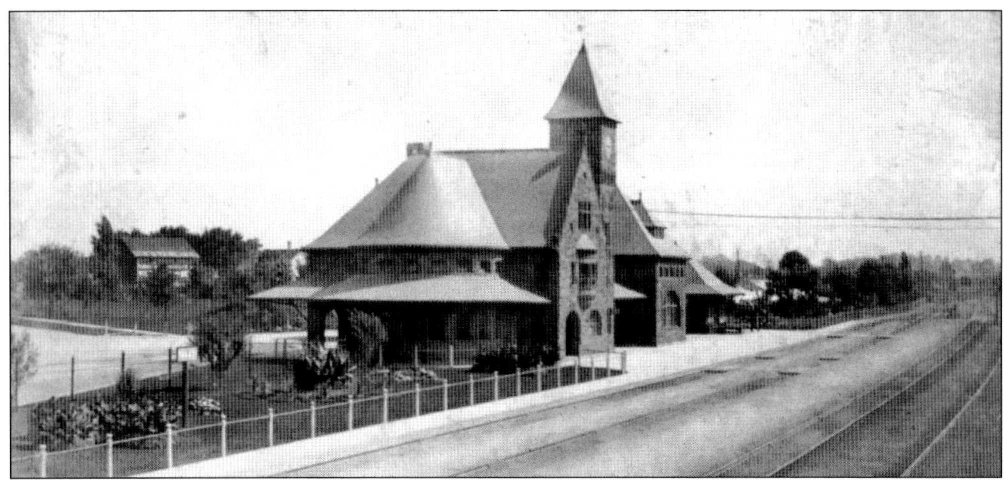

The Michigan Central built a grand new depot in 1892, designed to impress travelers en route to the 1893 World's Columbian Exposition in Chicago. The new railroad depot opened at Fifth and Dey Streets in 1892. The sandstone, Romanesque-style building featured a tall clock tower. After decades of the older, rundown depot, residents welcomed the new structure. Several major movies have been filmed at the depot.

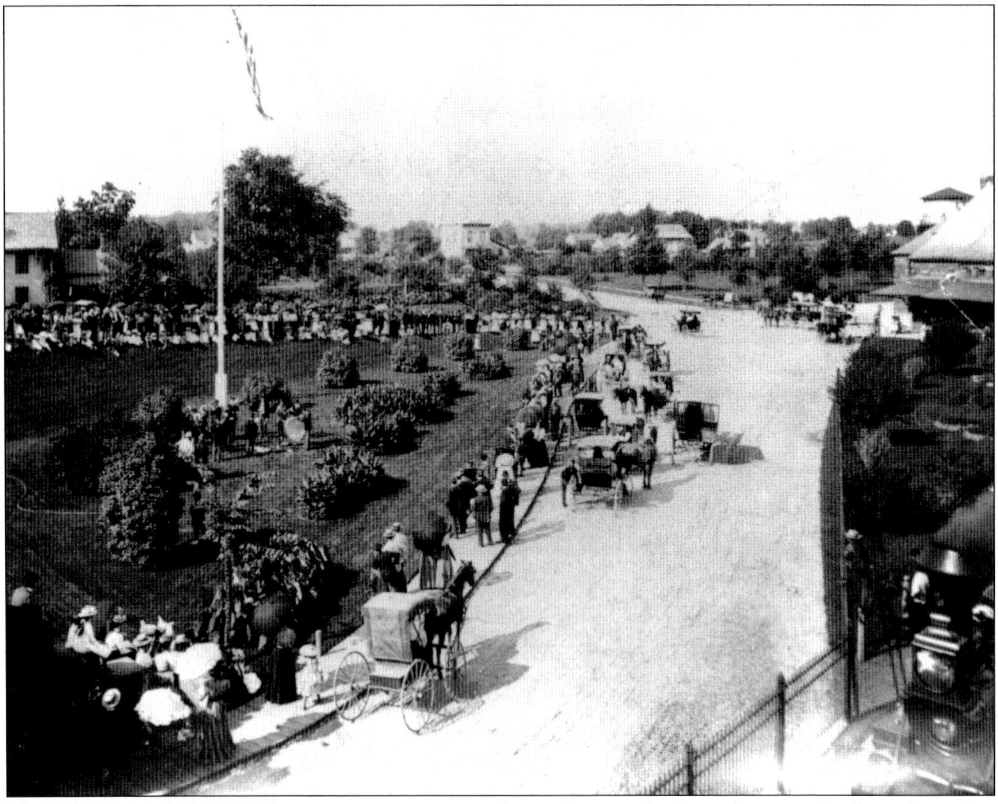

This photograph shows a flag-raising ceremony taking place at the depot's grounds. Crowds gather to witness the event and listen to a band playing near the flagpole. A locomotive is visible in the bottom-right corner, and carriages transport people to and from the trains. Judging by the number of parasols, it was a sunny day.

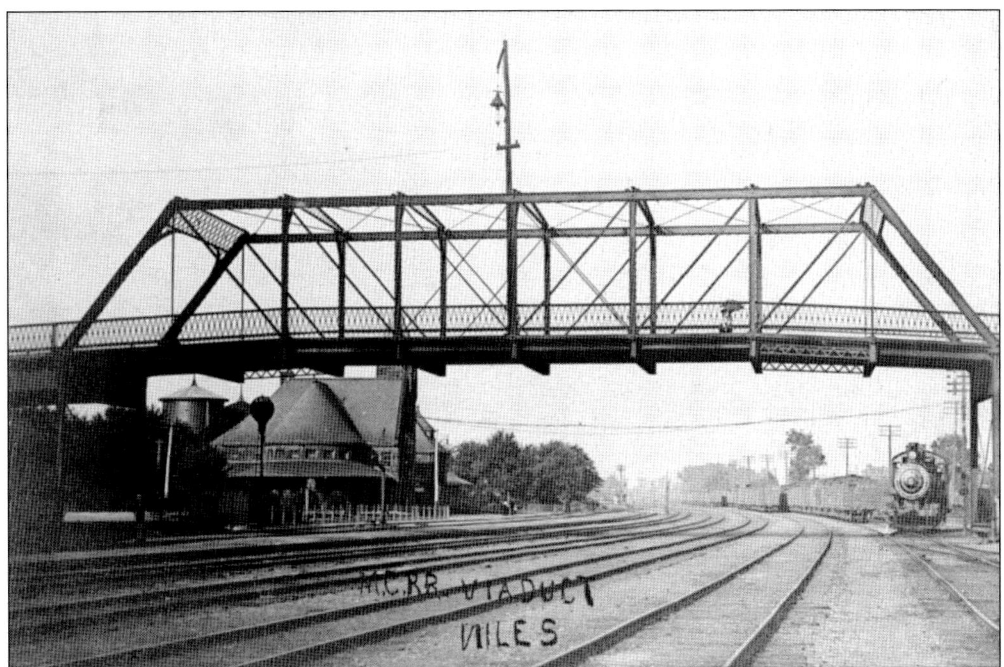

Along with the new depot, a viaduct was constructed to cross the railroad tracks at Fifth Street. Here, a locomotive is seen ready to pass underneath. After decades of foot traffic and later automobiles, the viaduct needed frequent repairs. Deemed a safety hazard, it was removed in 1977. Street crossing guards were installed instead of building a new bridge.

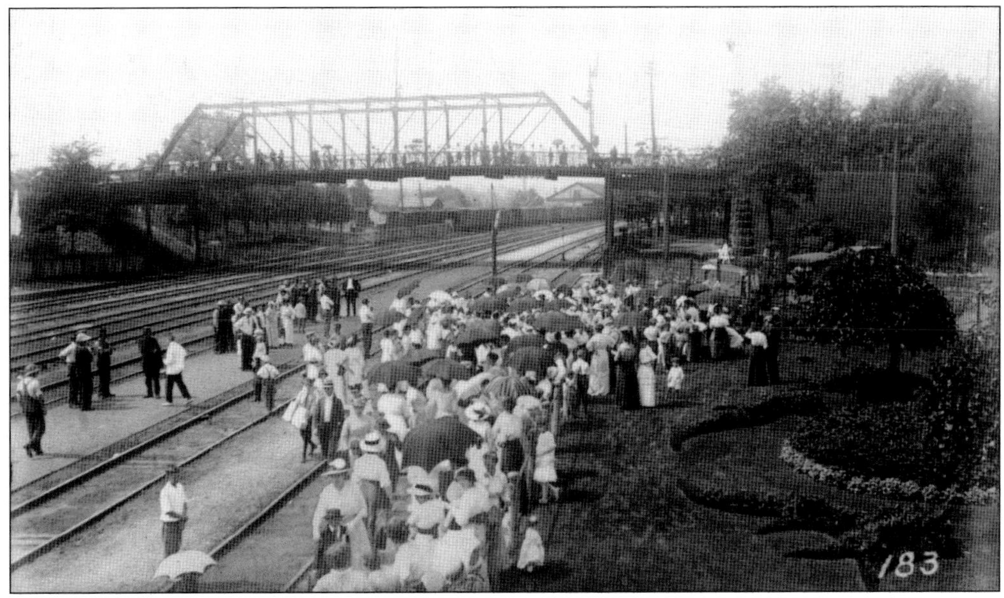

Nilesites greet children from Chicago who were sent out of the city to the country for "fresh air days." The Fresh Air Fund raised money promising "urban waifs" pure air, pure food, and pure fun. Area families hosted the children. Note the people waiting on the viaduct overlooking the tracks and the decorative gardens. (Courtesy of the Dowagiac Area History Museum.)

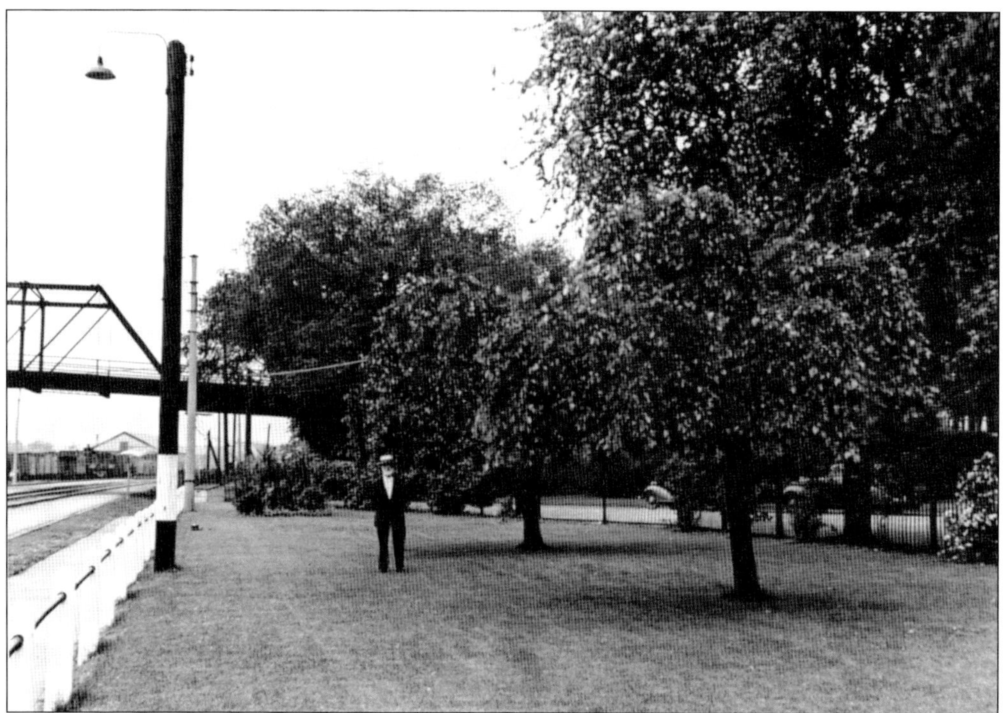

To enhance the impressive new railroad depot, master gardener John Gipner was hired to design the gardens. Greenhouses, colorful flower beds, shade trees, walkways, benches, and even a fish pond were added to the grounds. Niles earned the nickname of "Garden City" thanks to the efforts of John Gipner. He also started a custom of giving a flower to every lady passenger.

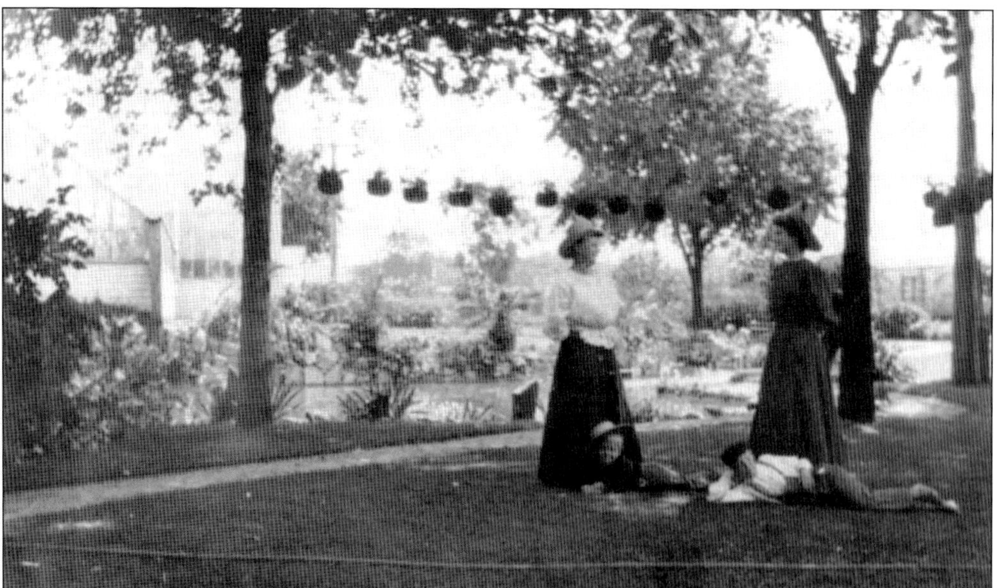

Railroad passengers and Niles residents alike enjoyed the depot gardens. Here, women and children relax near the pond. One of the garden's greenhouses is on the left, and a railroad car can be seen in the right rear area of the photograph. The greenhouse and pond are gone, but local garden clubs still plant flowers annually here.

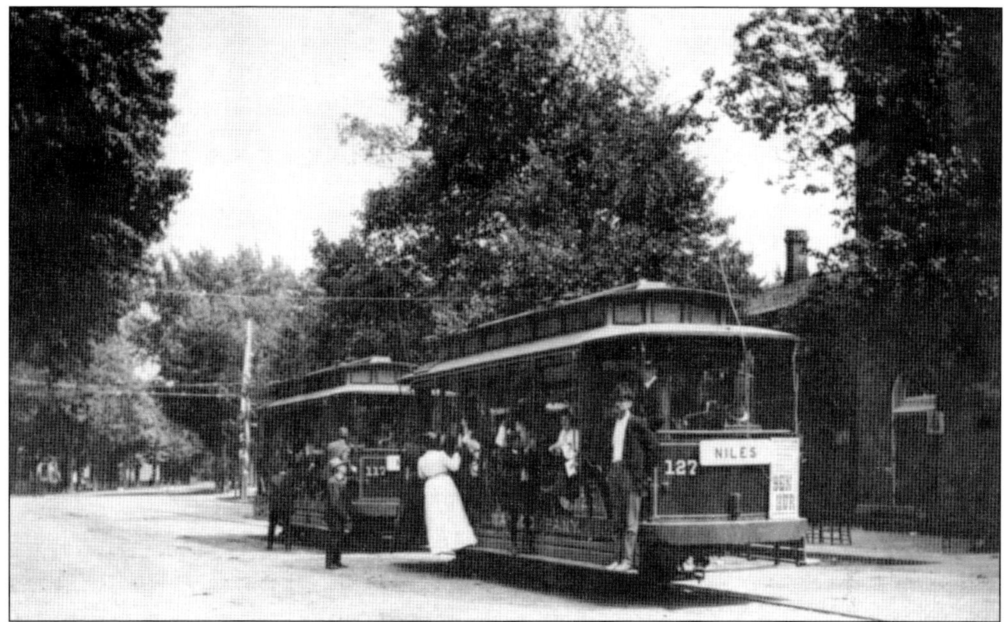

In the early 1900s, large railroads transported goods and people throughout the country, while interurban streetcar lines connected smaller towns. In 1902, the Niles City Council approved a franchise for the South Bend & Southern Michigan Railway Company. Interurban train service started from South Bend to Niles the following year. Eventually, the train traveled all the way north to St. Joseph, Michigan.

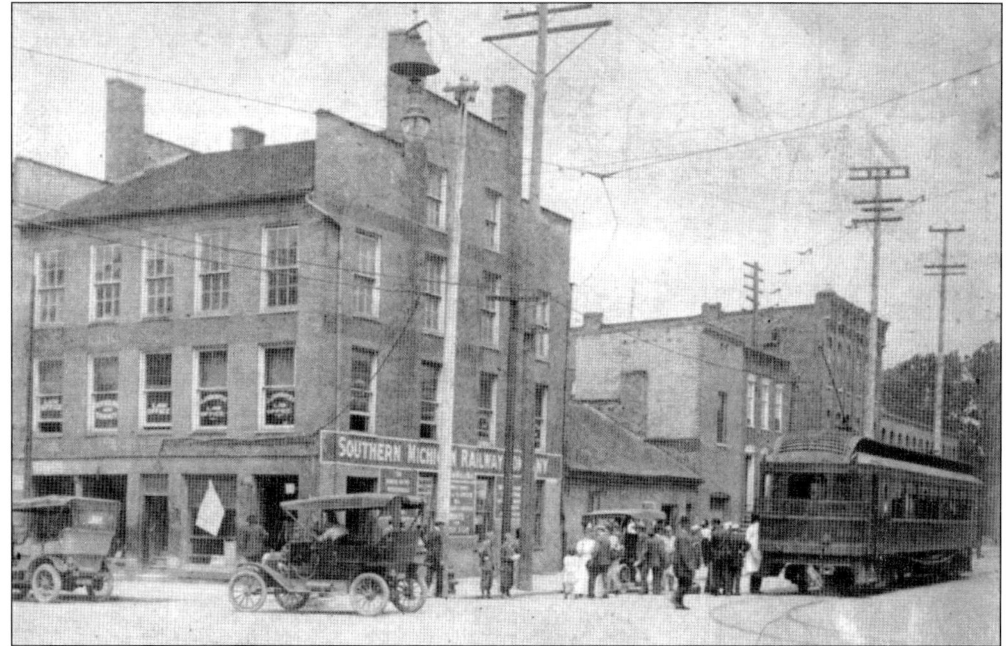

This photograph shows the interurban station at the northwest corner of Main and Second Streets. The sign on the building reads, "Southern Michigan Railway Company." An ad on the depot building promotes "Boats to Chicago," which riders could catch at the end of the railway's line in St. Joseph. Later, the depot moved two blocks north on Second Street.

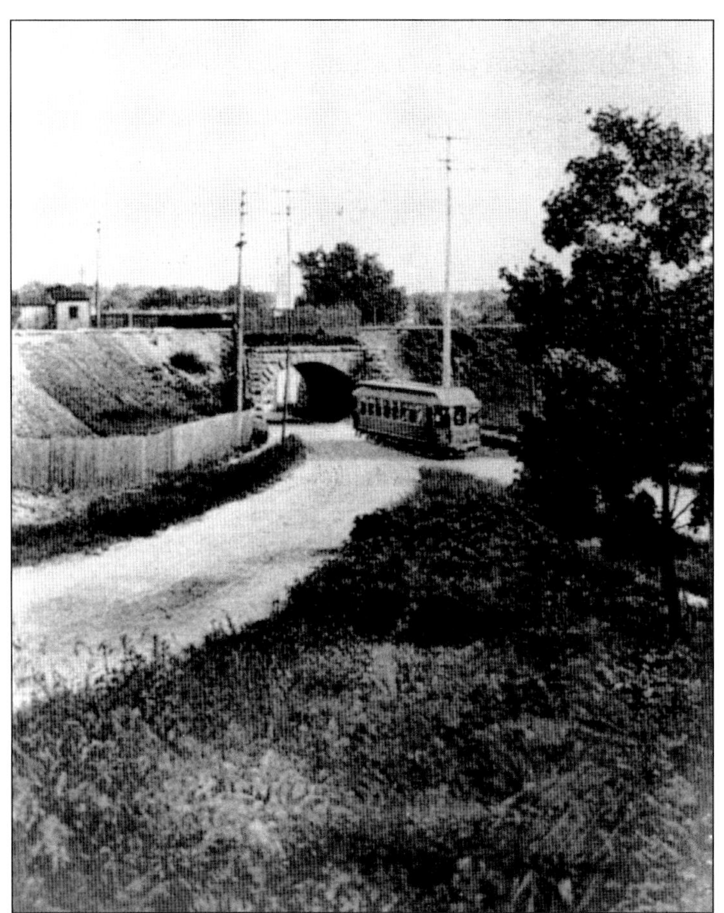

The Niles interurban railway passed under this stone archway, located at Second and Wayne Streets, traveling to and from Berrien Springs. The Michigan Central Railroad tracks ran on the bridge above the tunnel. Though the interurban tracks are long gone, the archway remains today as a reminder of the city's railroad history.

Two interurban trains make their way through downtown Niles. The train on the left is heading east on Main Street, destined for South Bend, while the other train travels north on Second Street to Berrien Springs. This photograph was taken in May 1906 and shows the corner of Main and Second Streets.

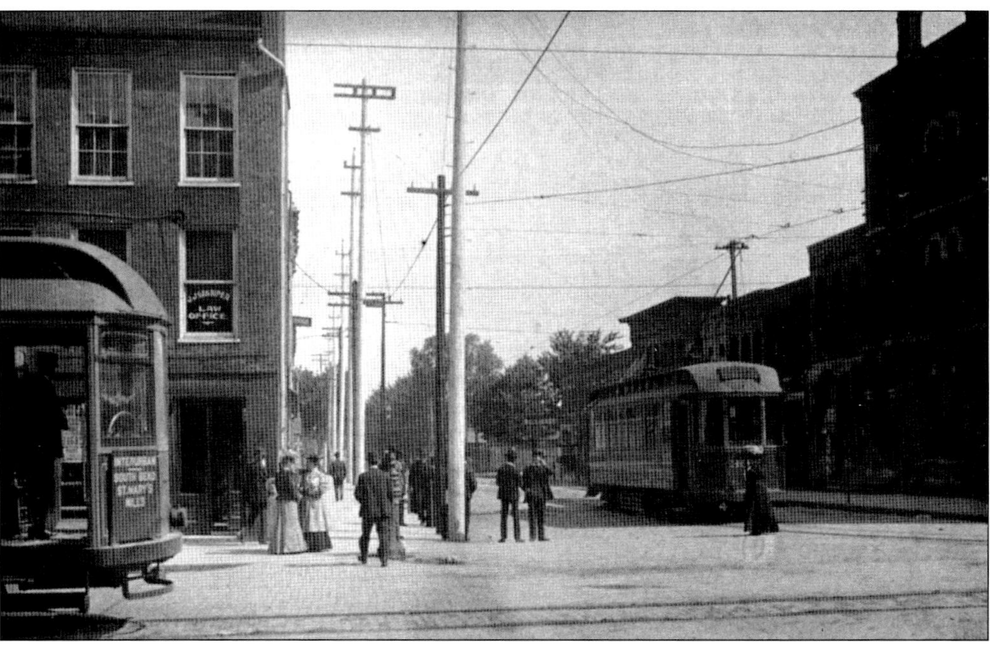

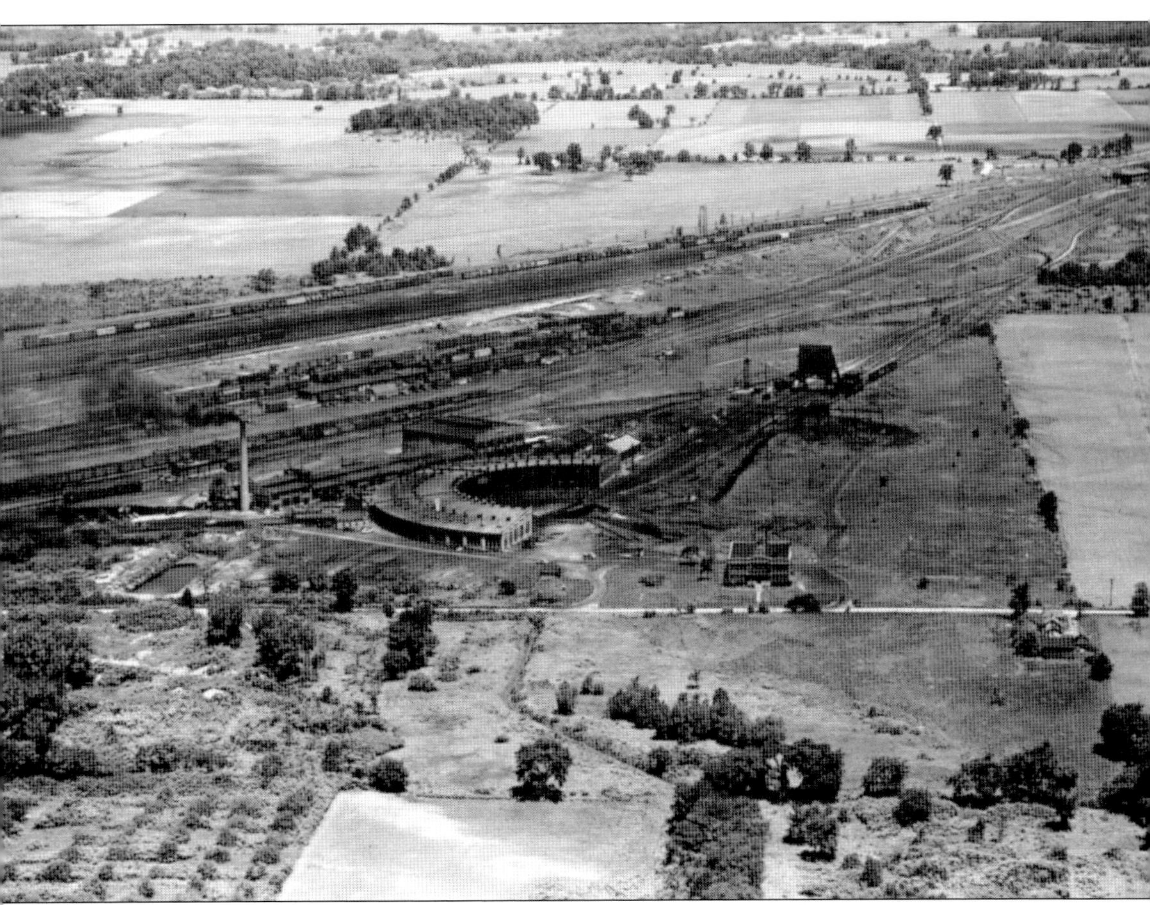

In 1919, the federal government made Niles the western division point for Michigan Central's freight operations. More than $3 million was invested in constructing classification yards to serve as a hub where trains were disassembled and regrouped for new travel. Expanded tracks, repair yards, coal stations, offices, a roundhouse, a turntable, and a hotel were built on Terminal Road east of downtown. More than 800 people were employed in the project, which was touted as "the Biggest Christmas Present Niles ever received."

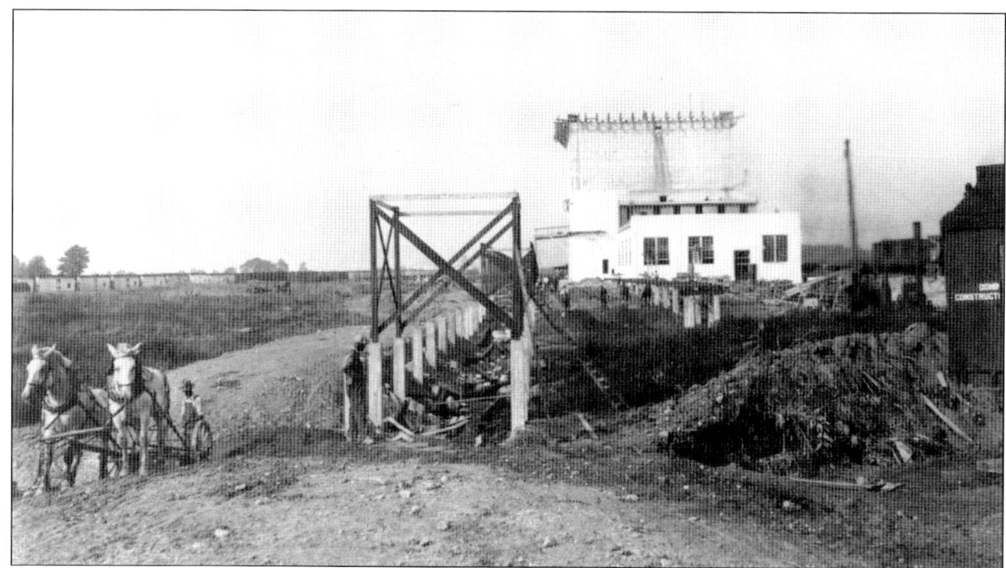

Niles railroad classification yards under construction are seen in this image. The structure in the middle will become the roundhouse turntable. The yards were one of the largest federal civilian projects to take place while World War I was raging. Horses on the left assist while train cars are visible in the background. (Courtesy of the Dowagiac Area History Museum).

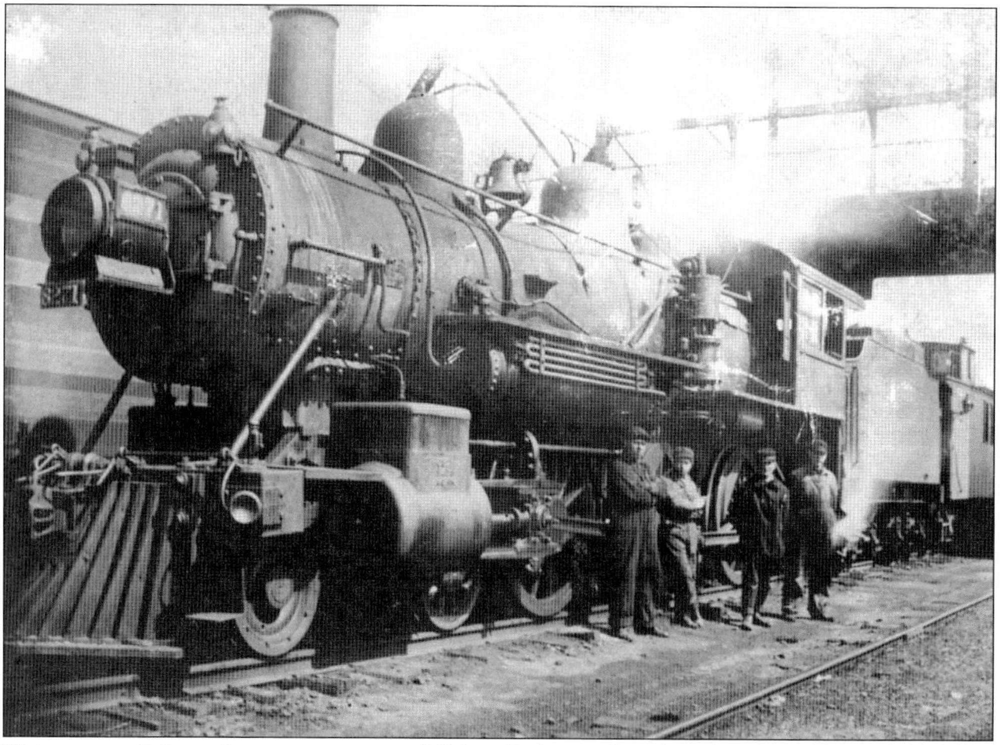

This postcard shows locomotive engine 8271 with four railroad workers. From left to right are Harry Saunders, engineer; William Smith, switchman; Edward McCoy, conductor; and Carl Dick, fireman. The photograph dates to 1915, and the caption on the back reads, "All set to help a freight train over the hill to Barron Lake." The airline branch of the railroad traveled to Barron Lake.

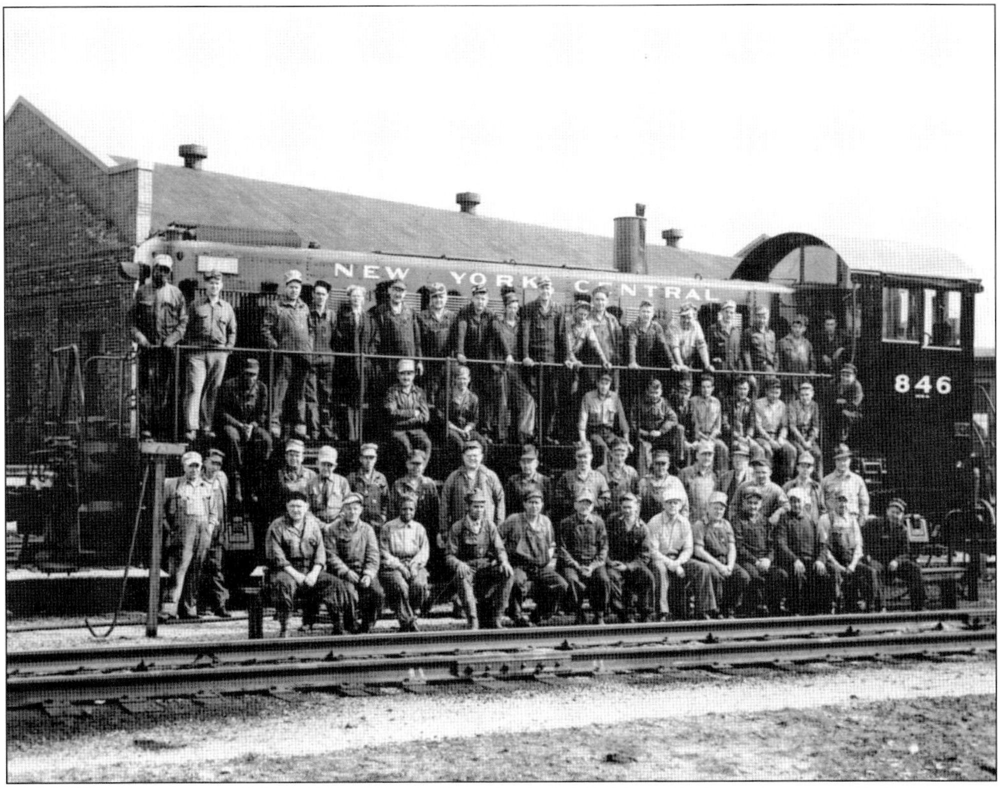

Railroad mergers and buyouts were common. Local lines had operated as part of the New York Central Railroad for many years. In the 1930s, NYCRR announced a name change for the Michigan Central Railroad and the Big Four. Though little changed at first, employees now sported NYCRR uniforms. NYCRR ran passenger trains in Niles until 1971.

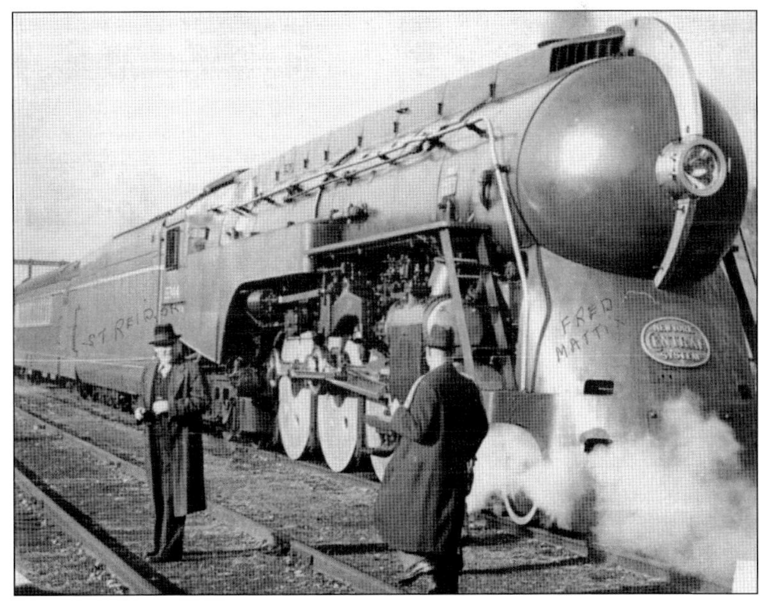

The railroad employed hundreds of people, from conductors to mechanics and station agents. Locomotive shops in Niles helped build the modern "bullet train," pictured here, for the New York Central Railroad. This photograph dates to 1938. Identified in the photograph are S.J. Reid, master mechanic (left), and Fred Mattix, train master (right).

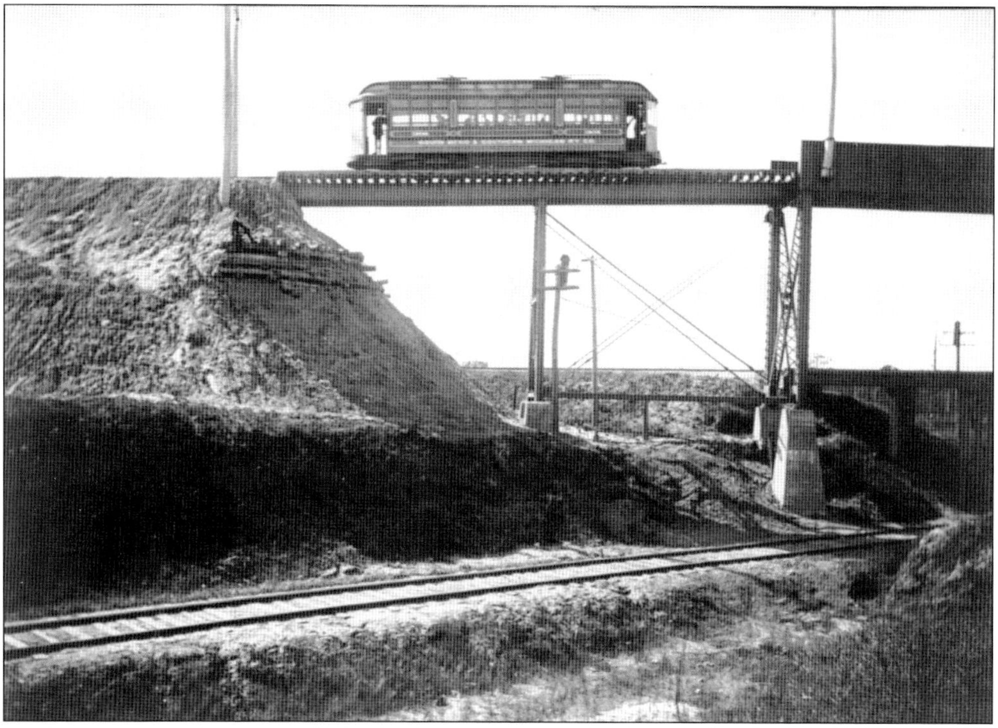

Once located near Fort and South Eleventh Streets, a trilevel bridge served three different railways. The electric interurban ran along the top track, as seen here; the South Bend branch of the MCRR traveled on the middle track, and the Big Four on the lower track. This bridge was likely the only trilevel railroad bridge in the nation. (Courtesy of the Dowagiac Area History Museum.)

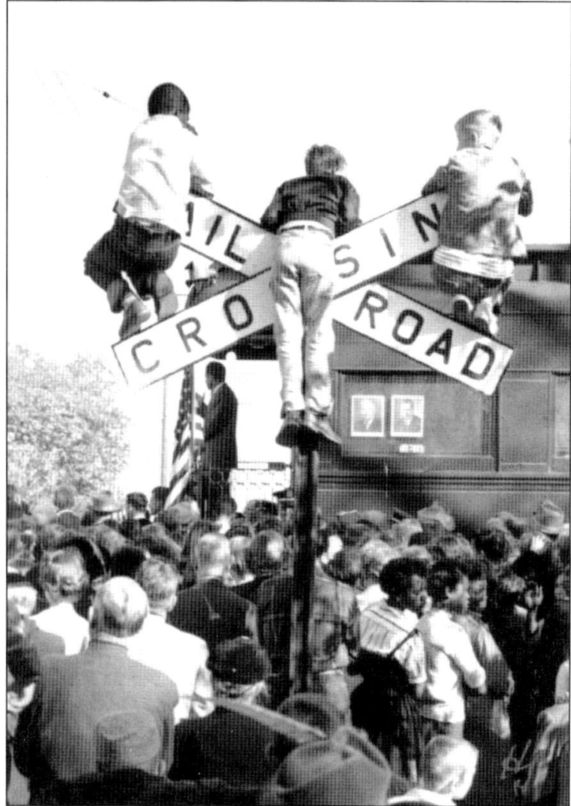

Niles photographer Harry Trescher captured this image of three boys trying to get a better view of Richard Nixon on a "whistle stop" tour as he campaigned for vice president in 1956. With its extensive railroad history and connections to major cities, including Chicago and Detroit, Niles was a popular stop for campaigning politicians.

Three

From Village to City

Through the 19th century, the village of Niles grew into a thriving city. By 1833, the Chicago Road, connecting Detroit and Chicago, passed through Niles, bringing scores of settlers. Bridges soon spanned the St. Joseph River, while dams harnessed its power to drive industrial growth. A business district developed along the riverfront and on Main Street.

The 1840 Niles Business Directory listed grocery and dry goods stores, clothing and shoe shops, a blacksmith, druggist, tavern, hotel, and more. Manufacturers produced furniture, copper, tin, sheet iron, and hardware for sale across the growing region. Teachers, doctors, newspaper publishers, and a circuit court judge were active in Niles before Michigan became a state in 1837.

By 1835, Niles had several hundred residents, and a village government was established. Just five years later, in 1840, the population had grown to 1,200. Village presidents led the community until 1859, when Elijah Lacey was elected the first mayor of the City of Niles. Local leaders met in various downtown buildings until 1933, when city hall moved into the opulent Chapin Mansion.

The Niles Police Department began as a one-man night watchman, operating from a small hut at Main and Second Streets. Independent fire companies and a citizen bucket brigade merged to form the Niles Fire Department. A public utility board brought water and electricity to the city, while business leaders and city officials worked together to pave roads, build bridges, and make other civic improvements.

Community institutions took root. Several local newspapers competed to bring news to Niles residents. A postmaster was appointed in 1830, with a post office following by 1835. The Ladies' Library Association led the effort to establish the city's first public library in 1904, funded by a grant from the Carnegie Foundation. A museum displayed fossils, relics, and curiosities. Schools and churches opened as families settled in the area, while doctors, dentists, and druggists provided health care. By the dawn of the 20th century, Niles, the largest city in Berrien County, was well on its way to becoming a regional industrial center.

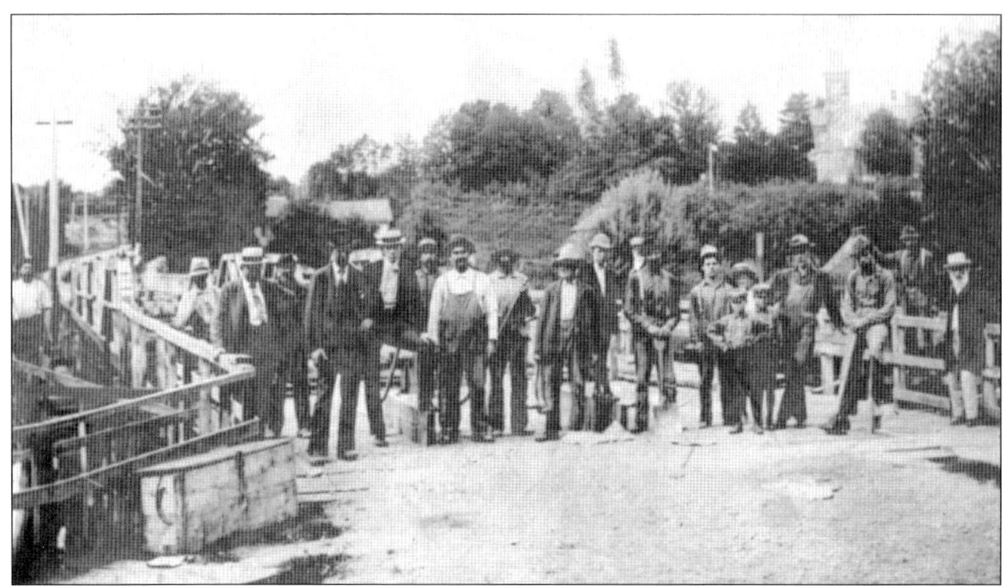

In 1835, a group of five men raised $2,500 to build a free bridge across the river at Broadway, replacing a rope crossing and ferry boat. Dr. John K. Finley served as chair and O.P. Lacey as secretary of the bridge committee. Completed in 1836, the new bridge allowed stagecoaches and heavily loaded wagons to pass. This image dates to sometime after 1870.

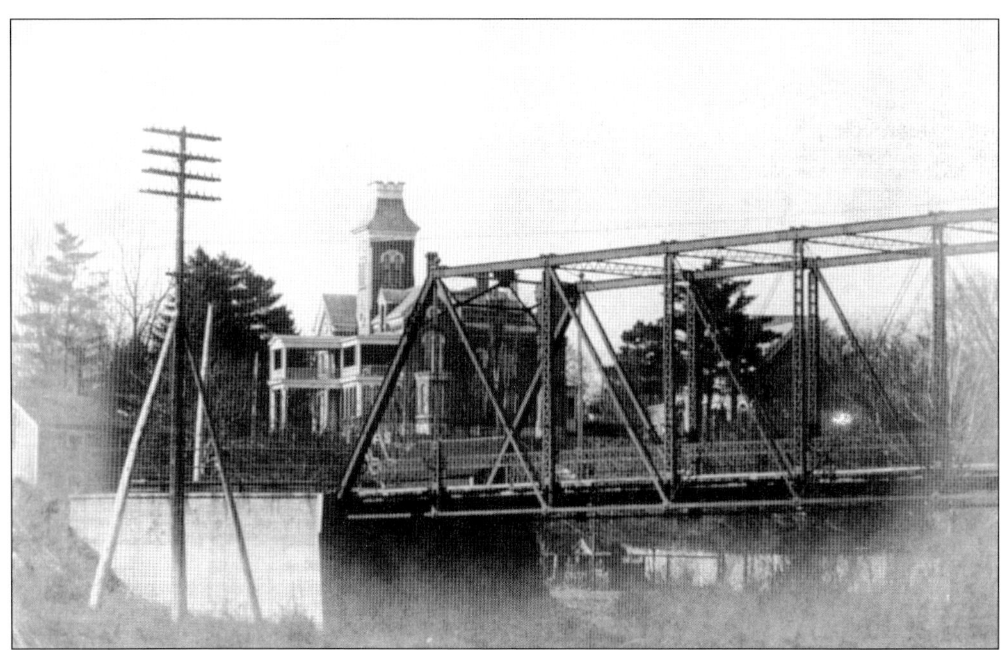

Although the Broadway Bridge was a free bridge, a sign noted that a fine would be assessed for "riding or driving faster than a walk." This image dates to the turn of the 20th century. The background shows the Castle Rest mansion on the river's western bank, which later became Pawating Hospital.

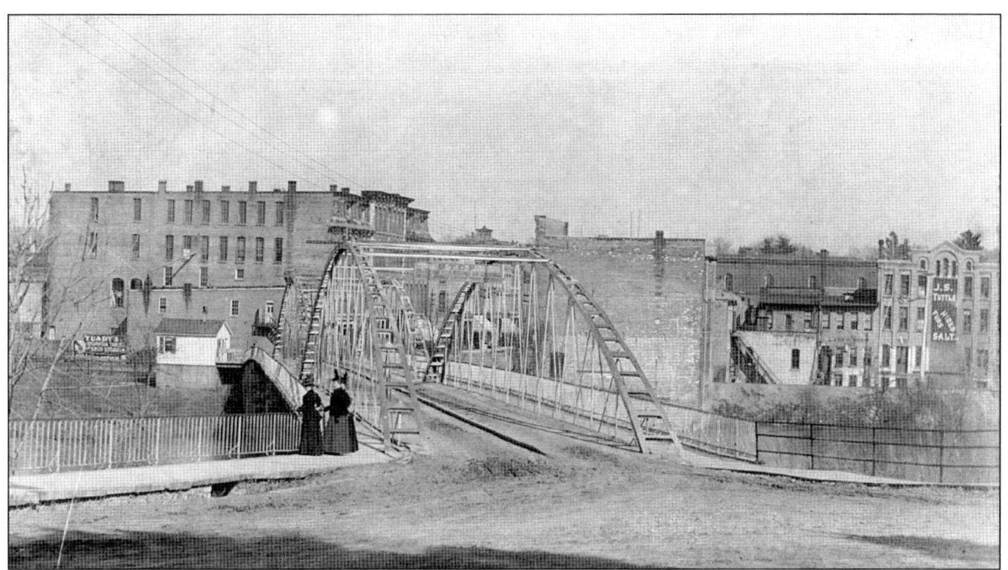

The first bridge to cross the St. Joseph River at Main Street was a wooden structure built in 1835. T.K. Green was elected Chairman of the Bridge Committee. This bridge required frequent repairs and was replaced by the iron bowstring truss bridge in 1868, shown here with a view from the west bank looking east.

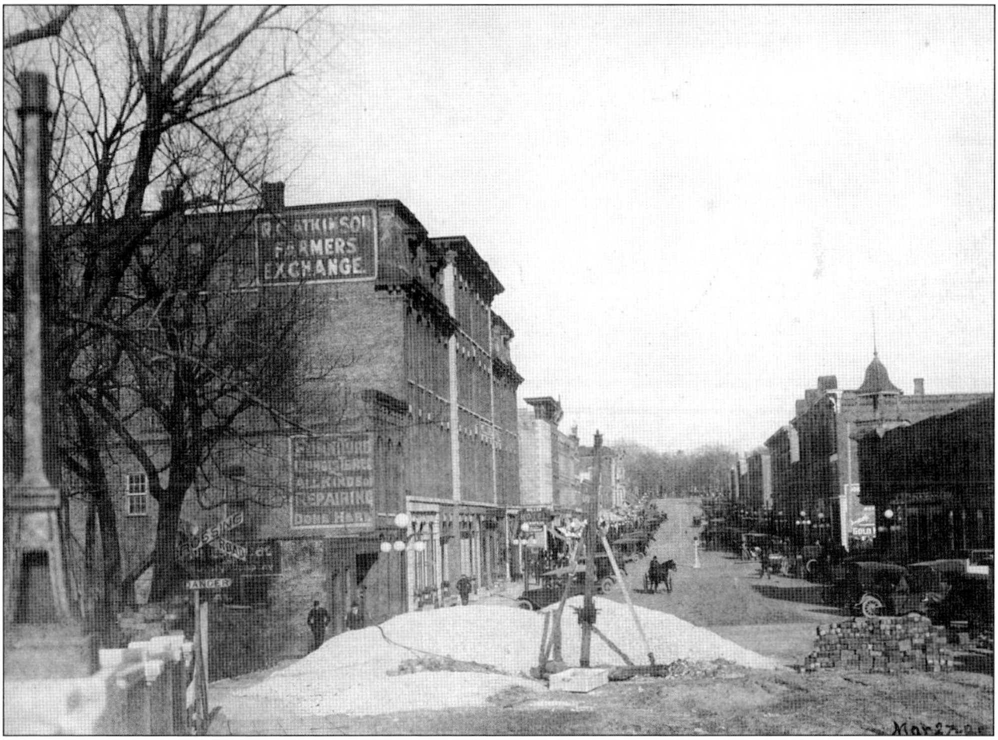

When the Main Street iron bridge fell into such poor condition that it had to be closed to traffic, plans commenced for a new concrete bridge. The council voted for the bridge construction at a cost of $75,000. The bridge opened to traffic in 1920. Note the crossing sign for the Big Four Railroad on the left.

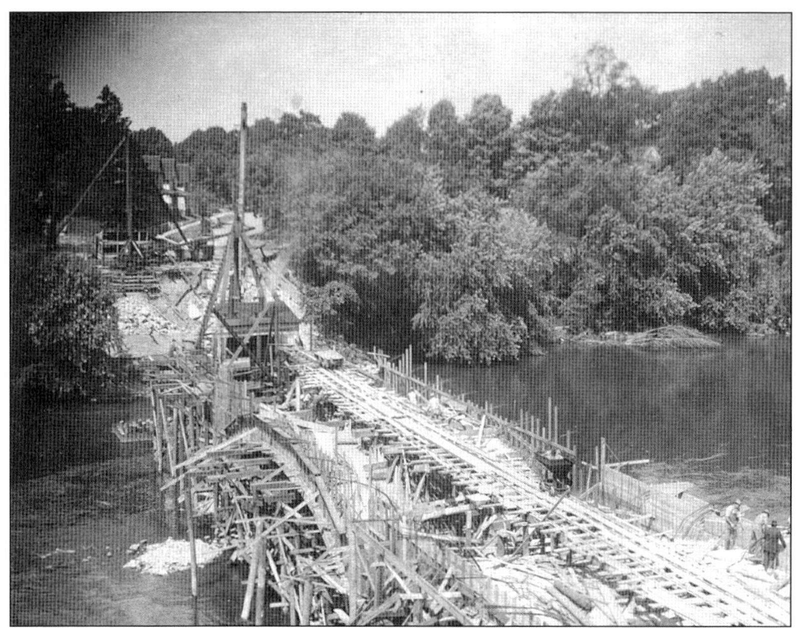

This photograph, taken looking westward from Niles City Hall when it was located downtown near the St. Joseph River, shows the construction of the new Main Street bridge in 1919. Some of the iron arches of the old bridge are still visible. Engineers are seen discussing plans on the right side of the structure.

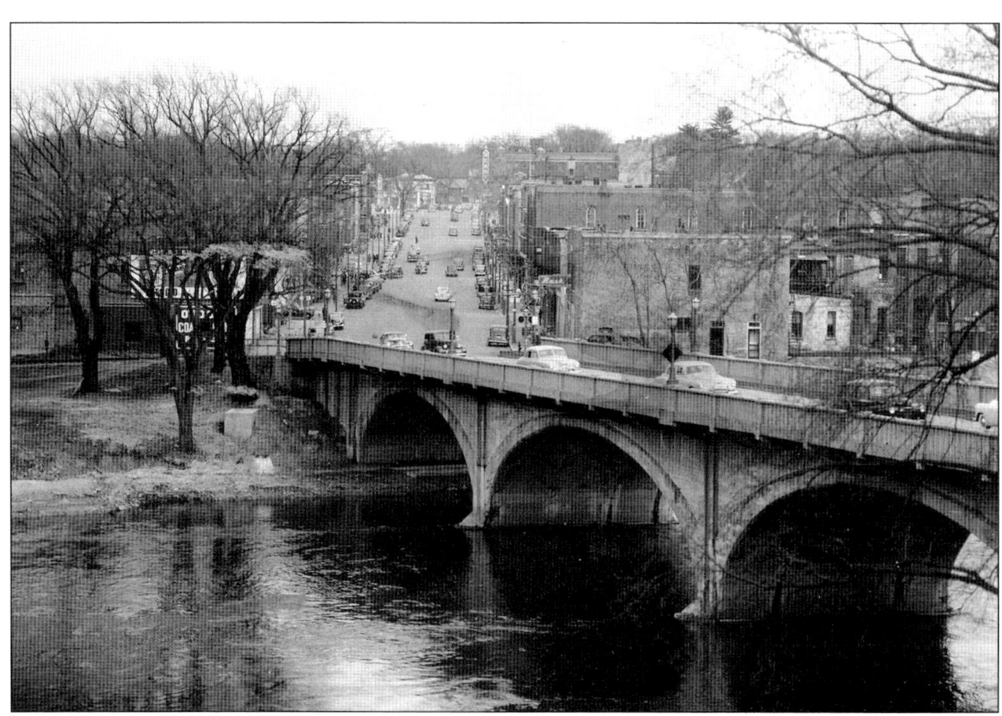

The 1920 Main Street concrete bridge lasted nearly a century. In 2015, it was replaced with a bridge designed to look like the 1860s iron structure. This is a view from the west side of the bridge taken in 1949. The large sign in the distance is for the Four Flags Hotel.

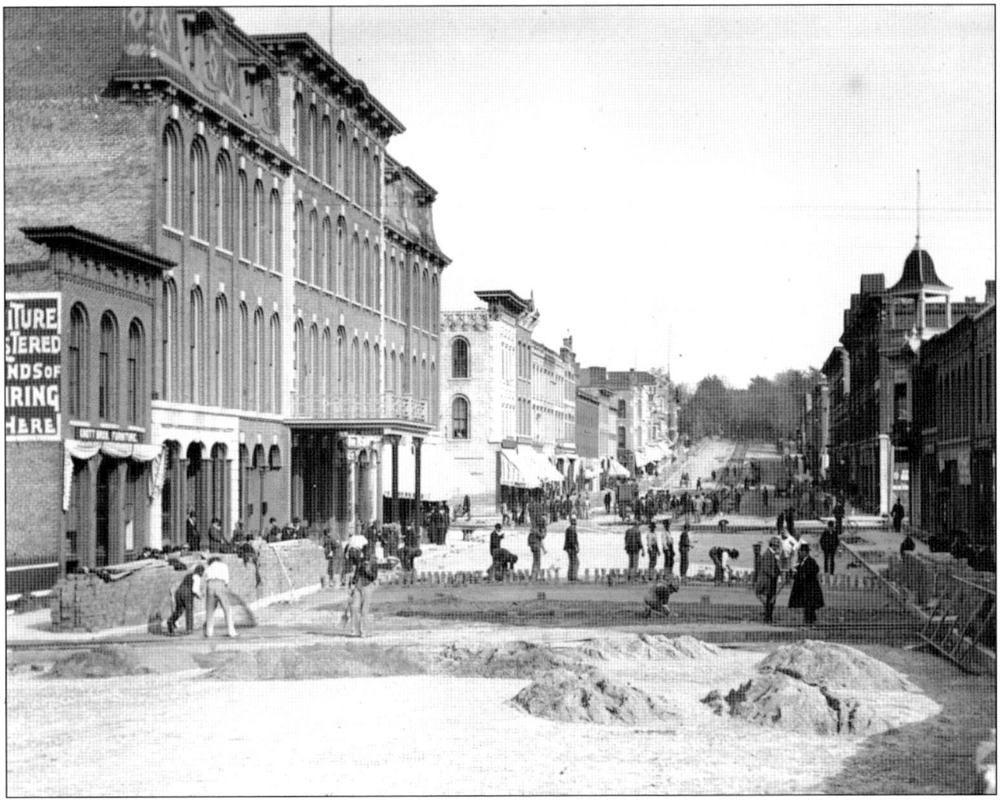

A crew lays brick on Main Street in 1902. A few months earlier, the city council passed a resolution to pave Main Street from the east end of the bridge to the intersection at Ninth Street. At this time, plans were also underway for the new interurban streetcar route, with the line planned for Main Street between Second and Fifth Streets.

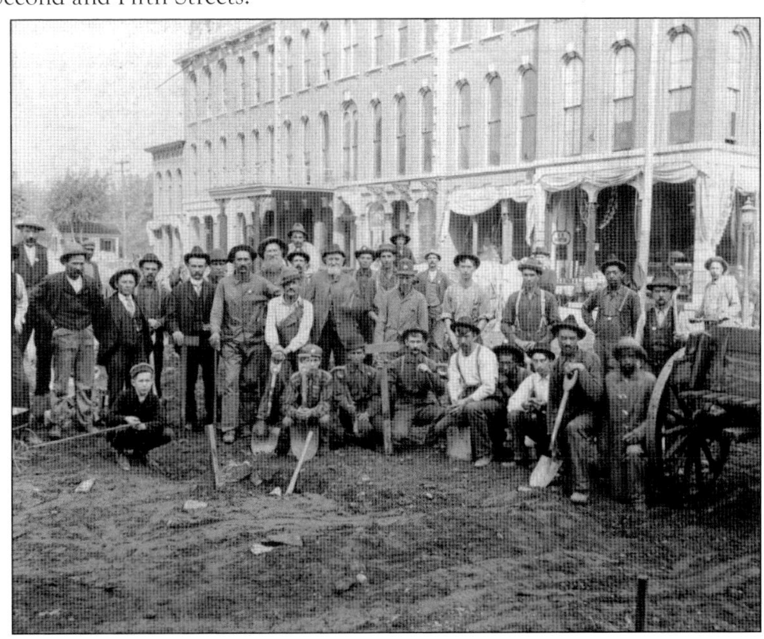

Members of the work crew take a break to pose for a photograph in front of the Michigan Inn. In November 1902, the *Niles Republican* reported that six miles of steel rails for the new interurban had arrived. Interurban service from Niles to South Bend began in August 1903.

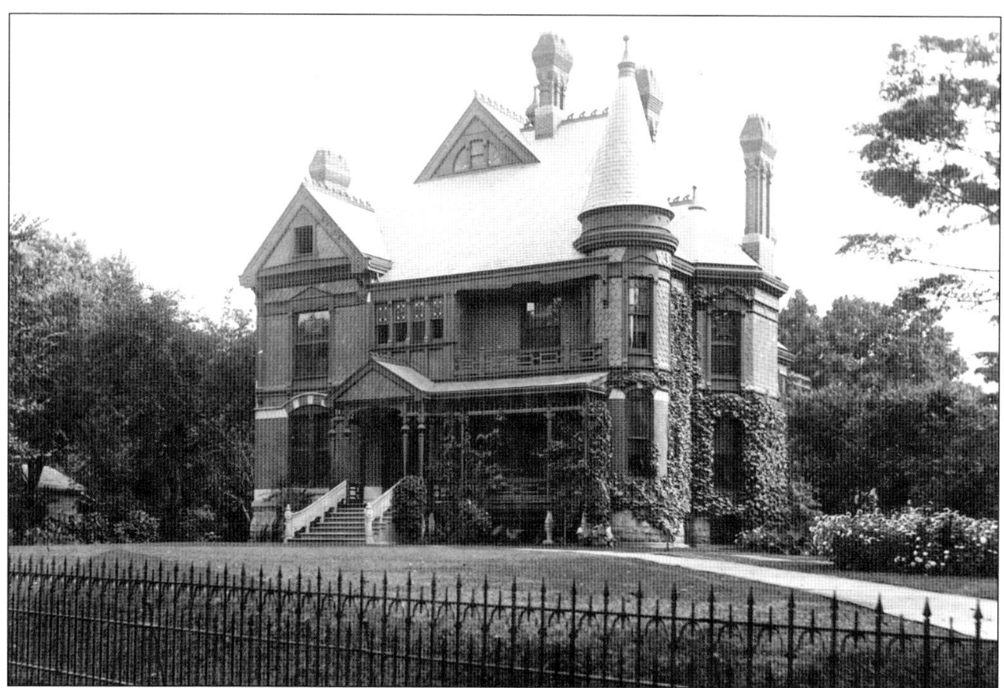

Henry and Ruby Chapin constructed their magnificent mansion between 1882 and 1884. The home stayed in the Chapin family until 1932, when their grandchildren sold the building to the City of Niles for a mere $300 with the condition that it be used for "civic purposes." The Chapin Mansion served as Niles City Hall from 1933 until 2012. The Niles History Center now operates the mansion as a historic residence while emphasizing its civic history.

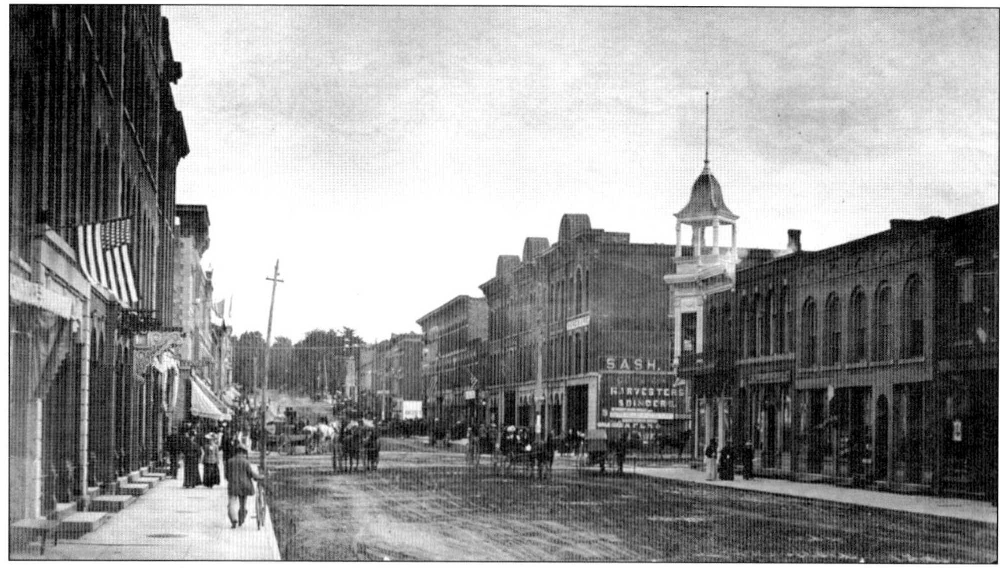

Here is a view of Main Street, looking east, on September 19, 1901. In the background, a crowd leaves the Niles Opera House after a memorial service for Pres. William McKinley, who was assassinated on September 14. According to newspapers, every business and industry closed for the service, and the hall was filled to capacity to honor the late president. (Courtesy of the Dowagiac Area History Museum.)

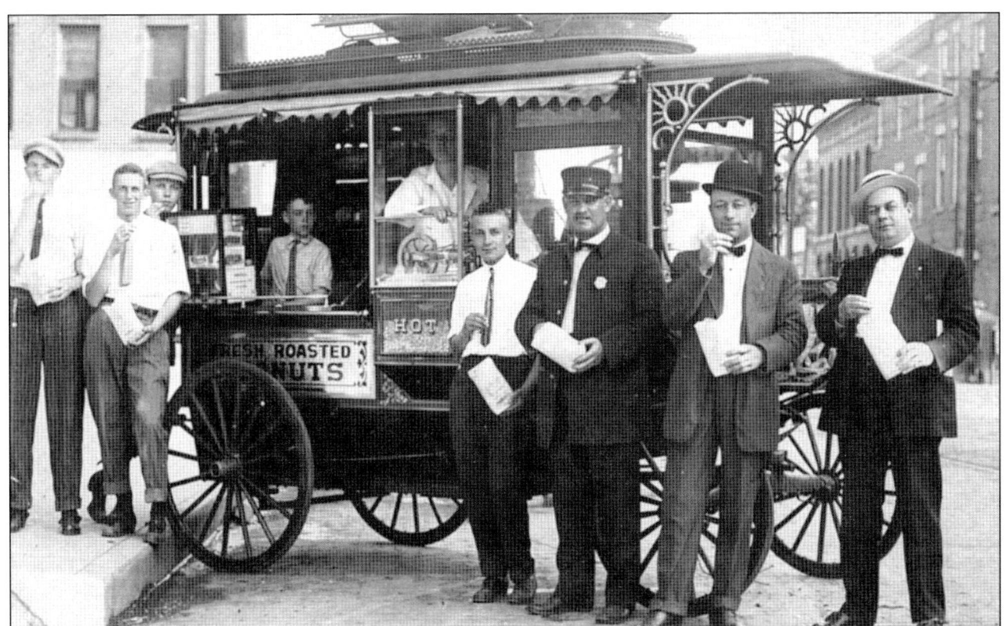

The 1851 Village of Niles Charter created the office of town marshal. The marshal worked with the ward constable to enforce laws. The Niles Police Department traces its origins to 1891, when the mayor appointed George "Dude" Francis to serve as the night watchman. Here, Dude Francis (center, in uniform) takes a break with Niles citizens to enjoy fresh roasted nuts from a street cart.

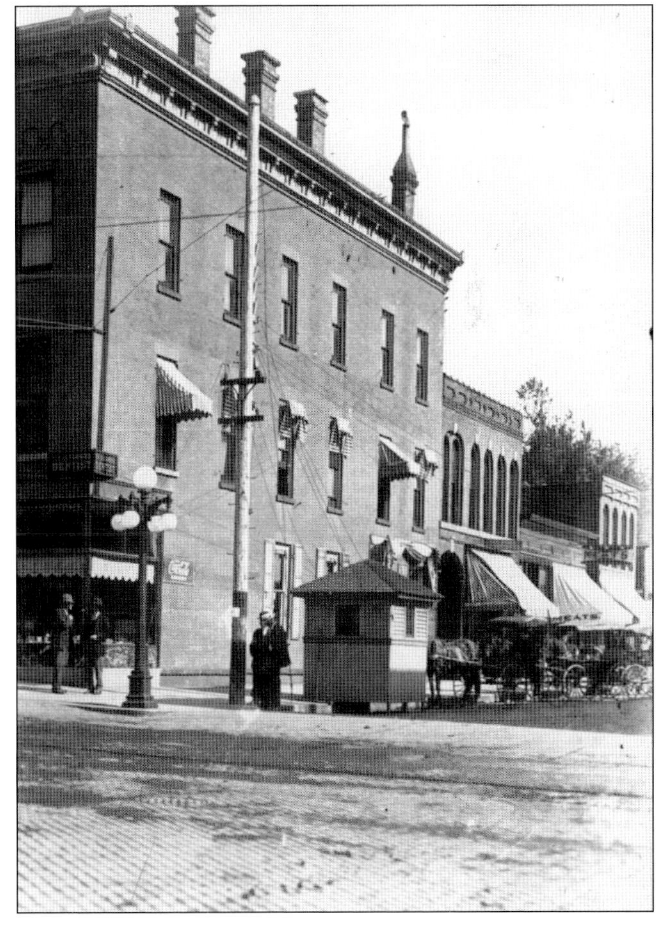

Night watchman Dude Francis operated from the small hut seen here, located on Second and Main Streets. Serving as the first police station, the hut was moved to Fifth and Wayne Streets in 1926 for use by patrolman Lawrence Elder. In 1955, the building was relocated again to the FOP (Fraternal Order of Police) Youth Park.

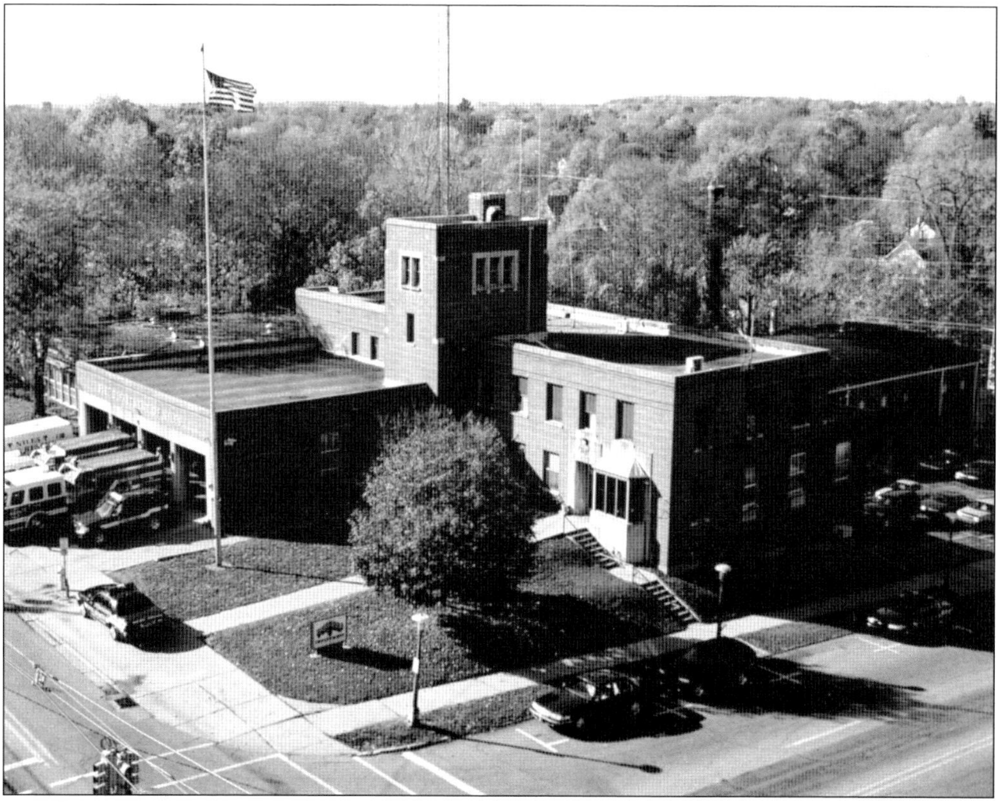

A Public Safety Building was erected in 1939 on the southwest corner of Third and Broadway Streets to house both the police and fire departments. This building was used until 2004 and was razed in 2015. Prior to this, the police station was in the same location, operating from the former residence of Dr. Homer Carr. The city had purchased the Carr residence in 1926 for use by the police.

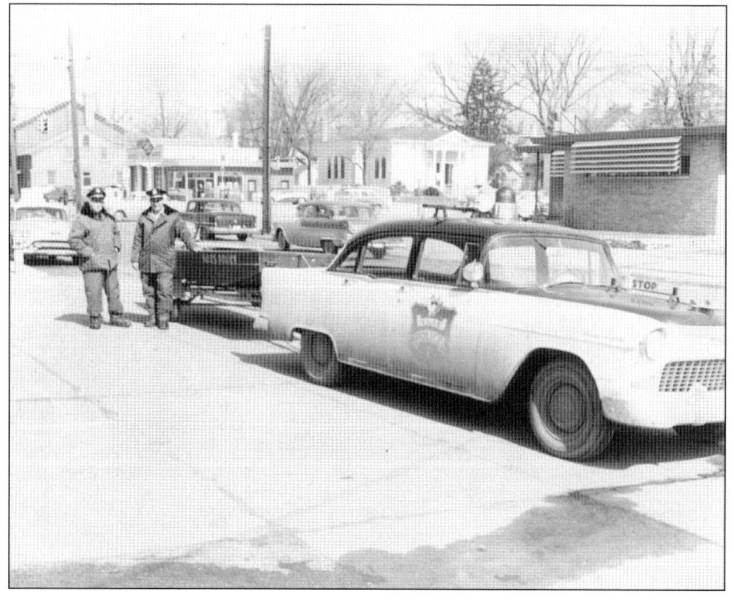

Two police officers stand with their patrol car at the intersection of Third and Broadway Streets near the Public Safety Building. Notice the police boat behind the car. With the St. Joseph River nearby, Niles police are often called to patrol the water. This photograph dates to 1955.

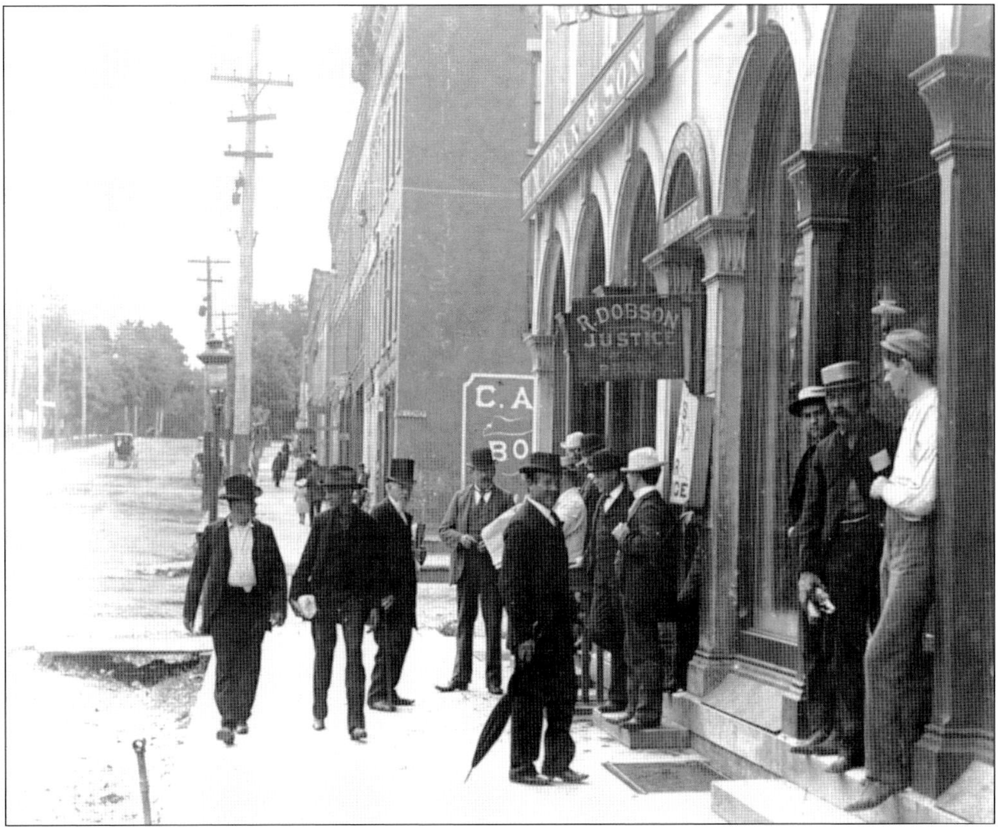

Men gather at the corner of Main and Second Streets in 1898, outside the office of the justice of the peace, Richard Dobson. Are they waiting for a marriage license? Not necessarily, as the justice of the peace was an elected official who tried limited civil cases in addition to performing marriages. The office was located over Henwood's Grocery Store on the 100 block of Main Street.

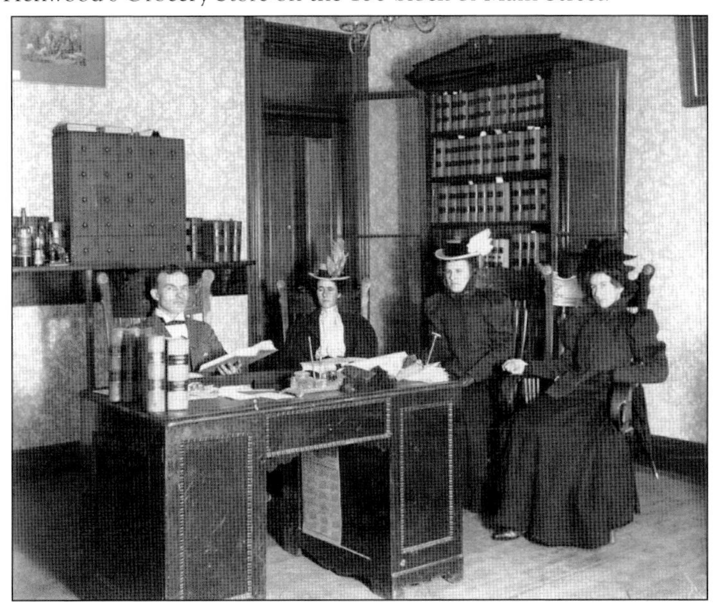

Nathaniel Bacon sits in his law office with Florence Bacon, Julia Gilbert, and Helen Woodruff. Born in Niles in 1867, Nathaniel Holley Bacon was part of the pioneer Bacon family that settled in the village of Niles during the 1830s. Like many members of the Bacon family, Nathaniel worked as a lawyer, with offices in the 200 block of Main Street.

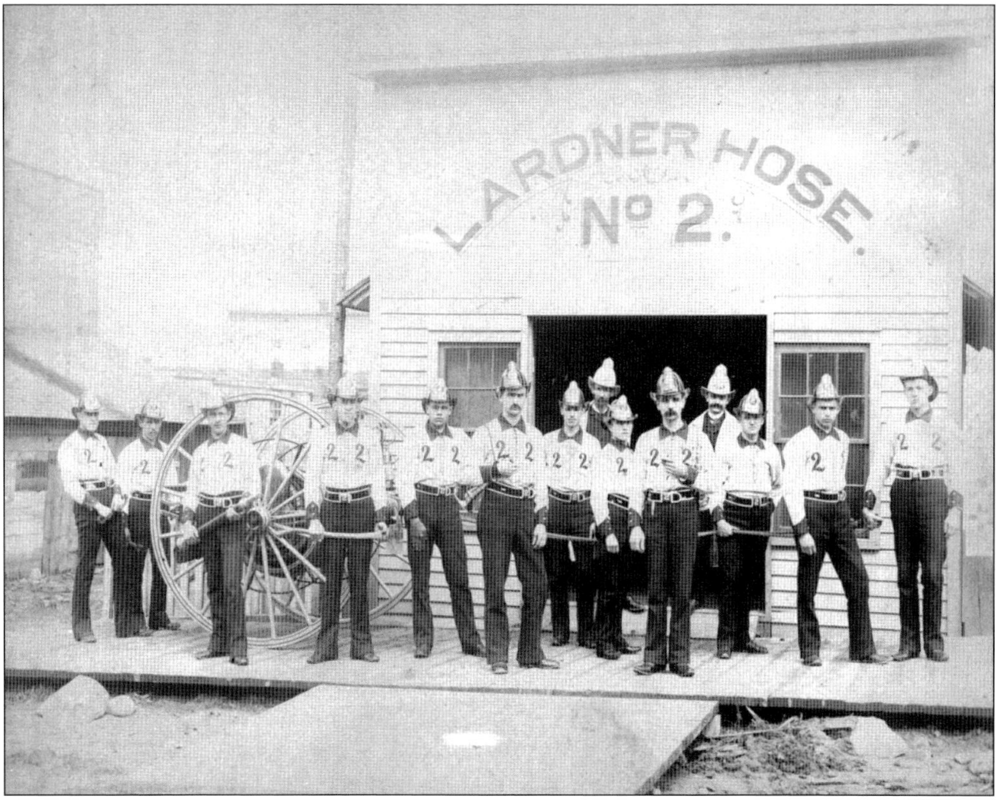

Early firefighters formed individual hose companies that were often named for their founders and outfitted with distinguishing uniforms and equipment. Pictured here in the 1880s is the Lardner Hose Company No. 2, named for the Lardner family, with their hose cart. William Lardner, brother of the famous author Ring Lardner, is on the far left.

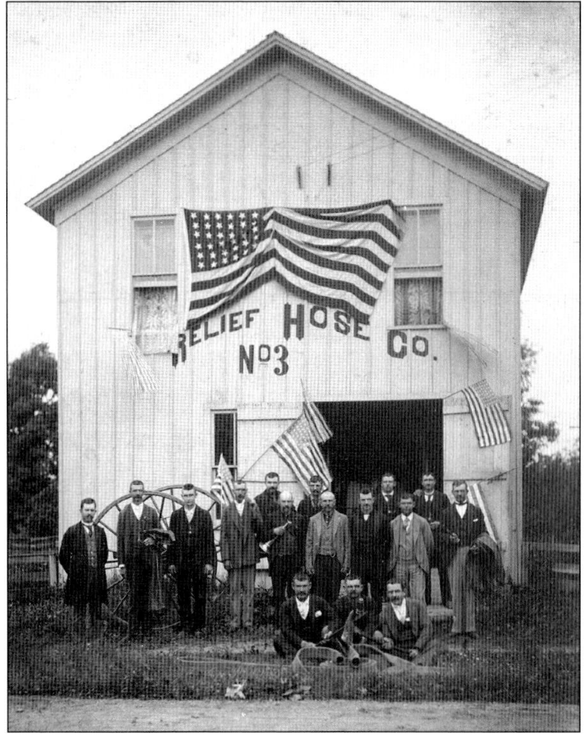

Some hose companies chose memorable monikers, like the Relief Hose Company No. 3, seen here with their hose cart and equipment in front of their station decked out with flags. In an unusual show of support for the company, some Nilesites smoked "No. 3" cigars, which were packaged in a special cigar box.

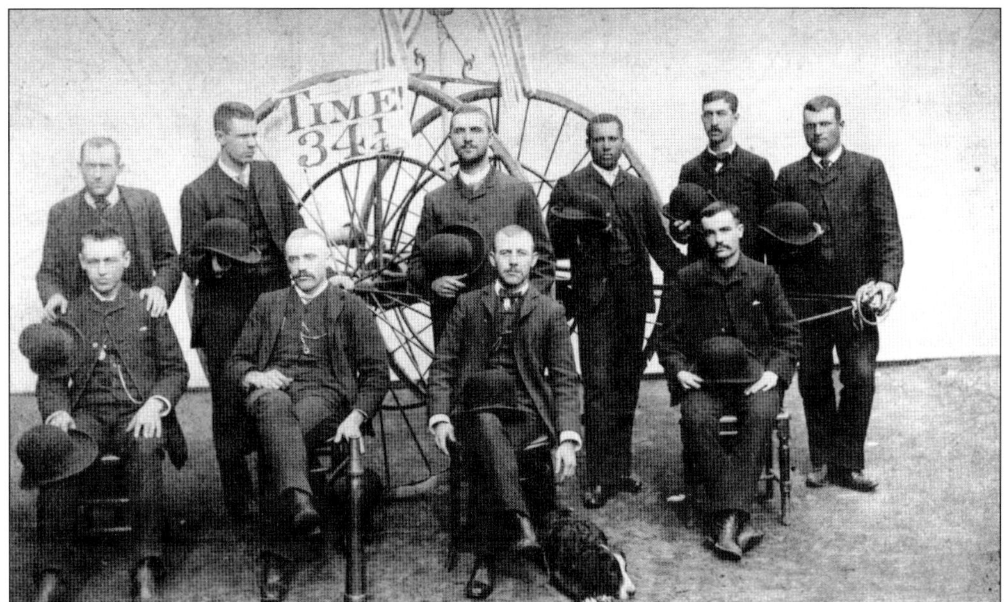

The Studebaker All Star Hose Team proudly poses after winning a contest. In 1882, the team laid and joined 100 feet of fire hose and sprayed water in 34.25 seconds, claiming a new "world record" for the time, as noted by the banner on their hose cart. Well-known eye doctor Fred Bonine led the team and is pictured standing in the second row, second from the left.

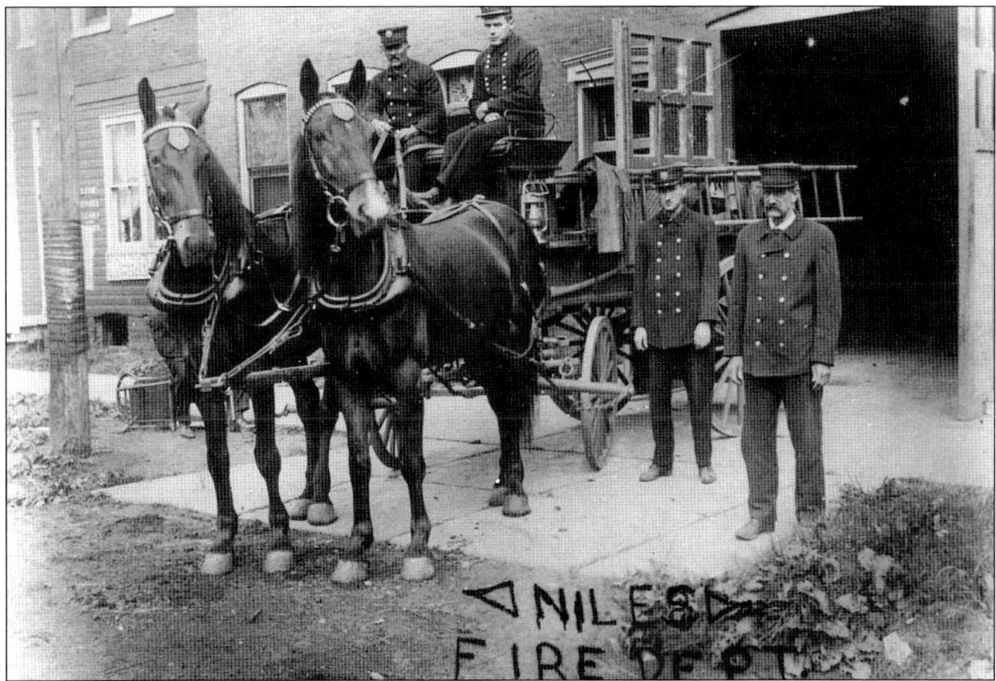

In the 1890s, construction started on a new Central Station in the 100 block of Sycamore Street. Central Station served as the fire department's home until 1939, when firefighters moved with the police into the new public safety building on Third and Broadway Streets. Central Station firefighters used horses to move and carry equipment.

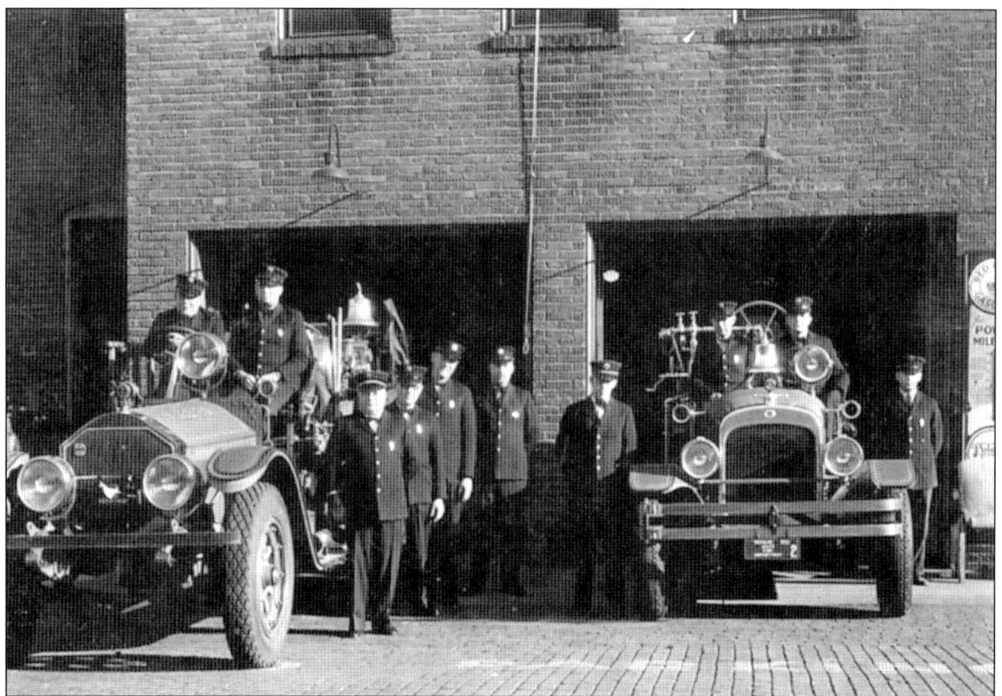

After the days of horse-pulled hose carts, the stables at Central Station were changed over to garages. The Niles Fire Department poses with their trucks outside of the station around the year 1932. At this time, the Niles fire chief was Clayton Collins. The building that housed Central Station was demolished in 1989.

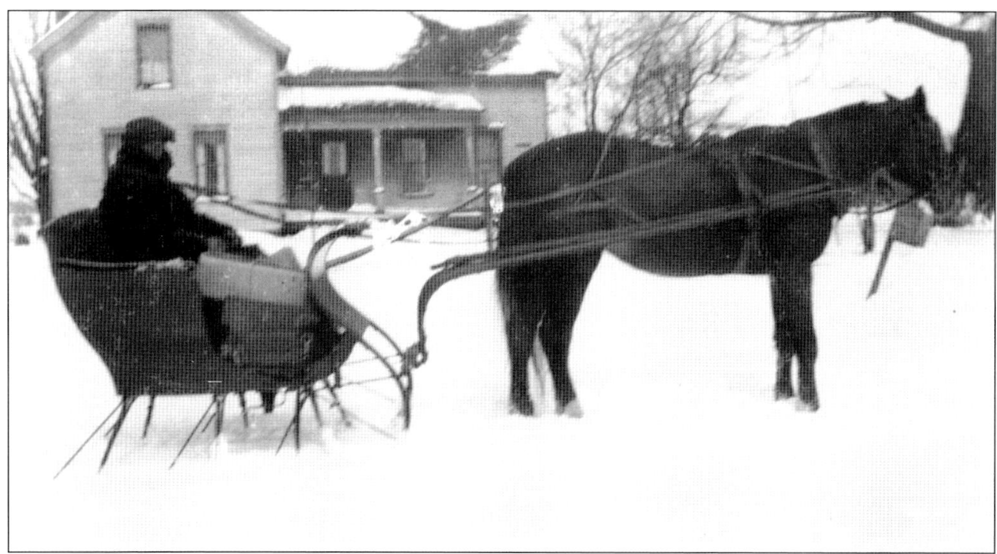

Bert Fisher delivers rural mail during the winter in a one-horse open sleigh around the year 1901. The Fishers had a farm three miles northeast of Niles. Bert Fisher worked as secretary for the Rural Mail Carriers' Association of Cass and Berrien Counties. He also served as a delegate to the State Convention of Letter Carriers. Bert delivered mail for more than 30 years.

Nilesites wait for the post office to open in 1898. Until 1910, the post office was located at the northwest corner of Third and Main Streets and served around 12,000 people each year. Four rail lines made Niles a postal hub. On the lower level was Calvin Wilson's Barber Shop. Wilson was an African American barber and active in local politics. His name is visible in the lower left corner.

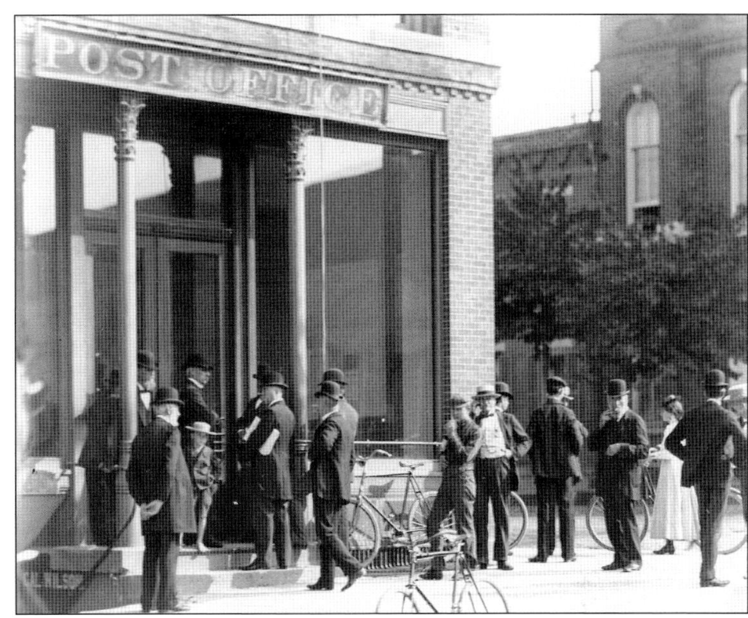

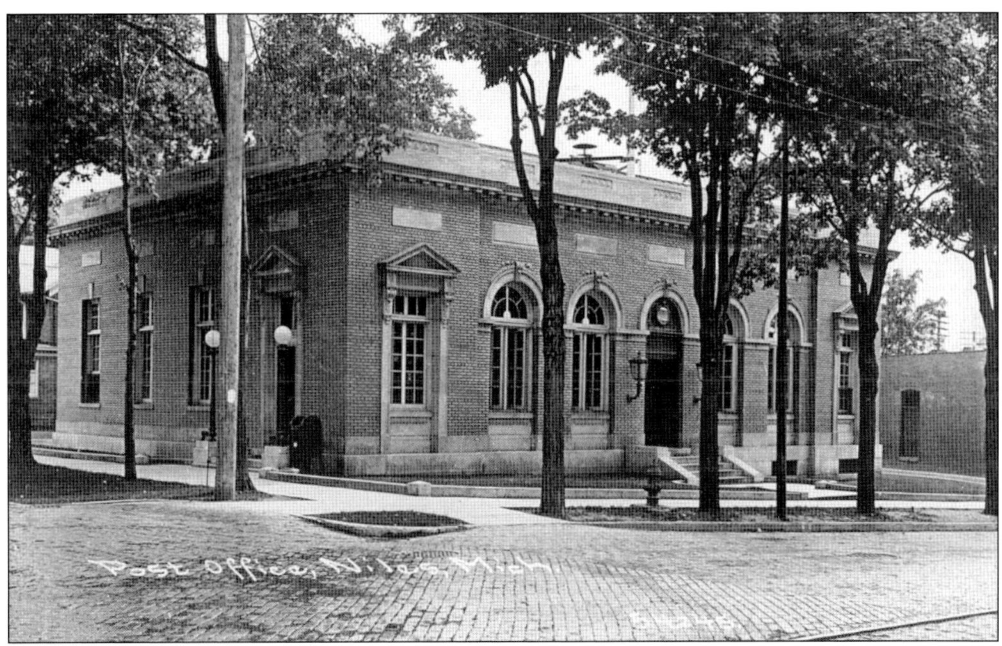

Niles's second post office opened in 1910 with Carmi R. Smith serving as postmaster. The postal service operated here until the 1980s, when a new post office was built at Ninth and Broadway Streets. The 1910 building later served as the home for the City of Niles Utilities Department. Today, the structure houses offices for the Post Office Apartments, the name hearkening back to its roots.

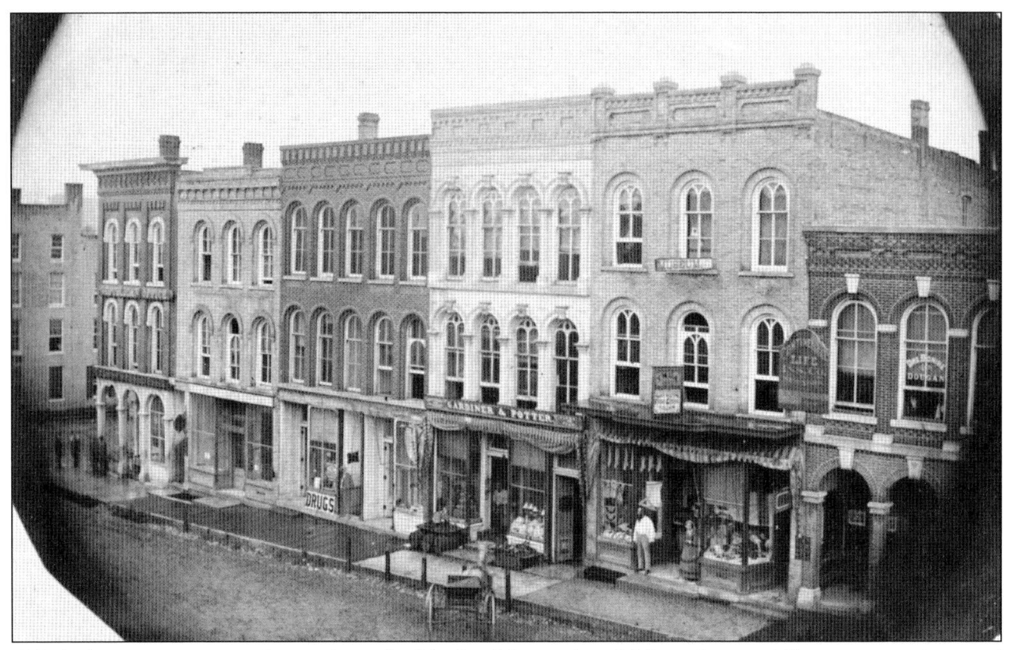

Niles's first museum was located on the block of Second and Main Streets. The museum featured the collections of Clement L. Barron, largely consisting of taxidermy and natural history specimens. The museum also displayed battlefield relics from the Civil War and items collected from the site of Fort St. Joseph. Admission cost 10¢. Dating to 1869, this photograph shows the museum sign visible on the third-story windows.

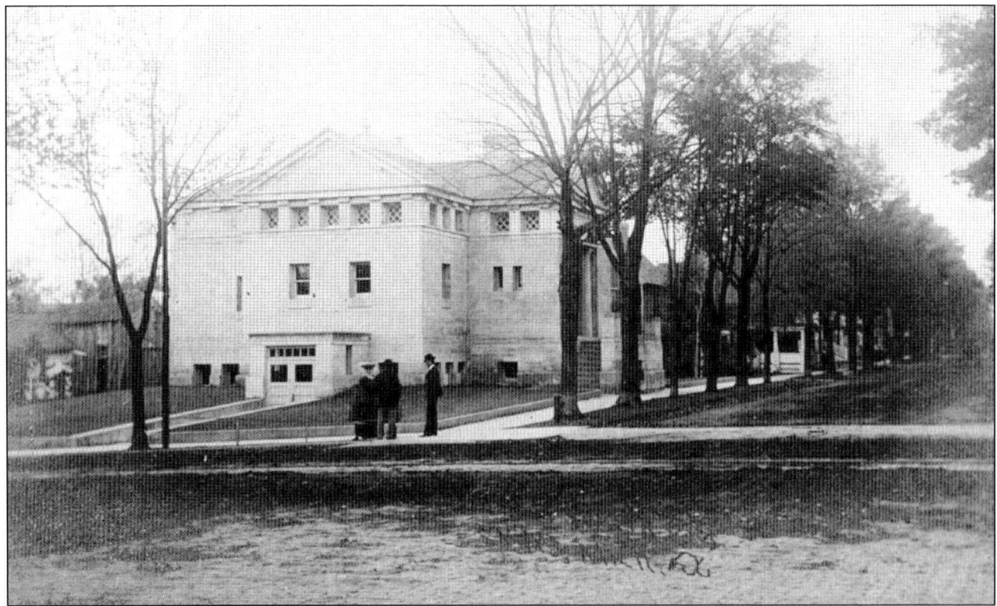

Niles's first public library opened in 1904, funded through a grant from industrialist Andrew Carnegie matched with city dollars, the Federation of Women's Clubs, and private donors. In 1910, the *Niles Daily Sun* boasted that Niles was one of only 40 cities with a Carnegie Library. After a new library opened, community groups, including the Four Flags Chamber of Commerce, used the building, which still stands at Fourth and Main Streets.

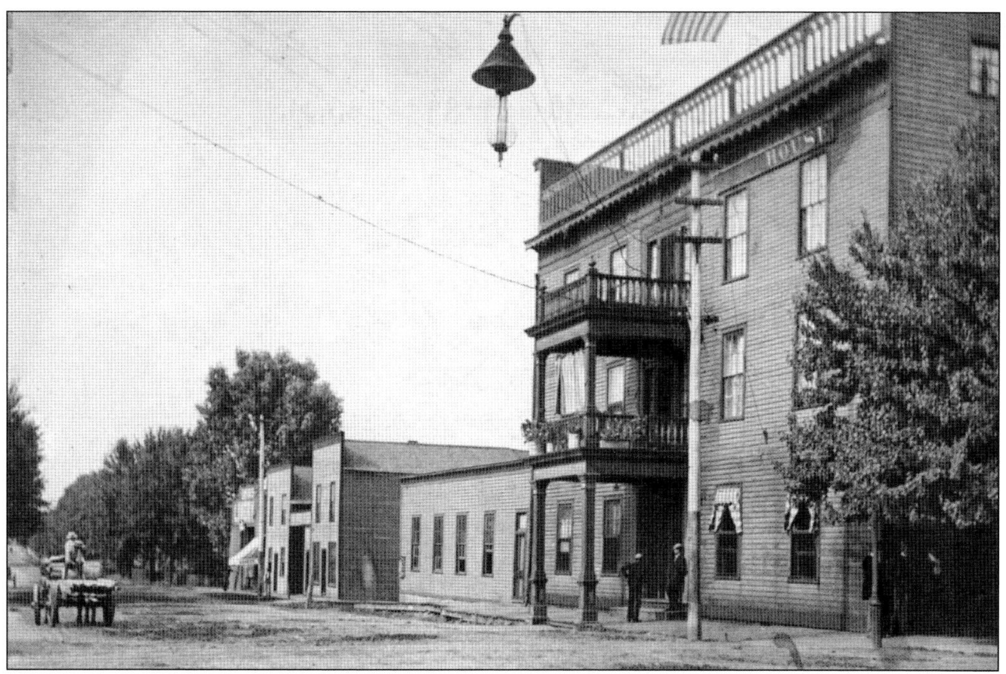

Built in 1852 as the Niles Exchange, this early hotel was renamed in 1857 to honor esteemed citizen, Judge John G. Bond. Considered Niles's finest hotel, the Bond House boasted a guest list of notable names. After several ownership changes, however, its quality dropped. Hoping for a new start, the hotel was renamed the Galt House in 1892.

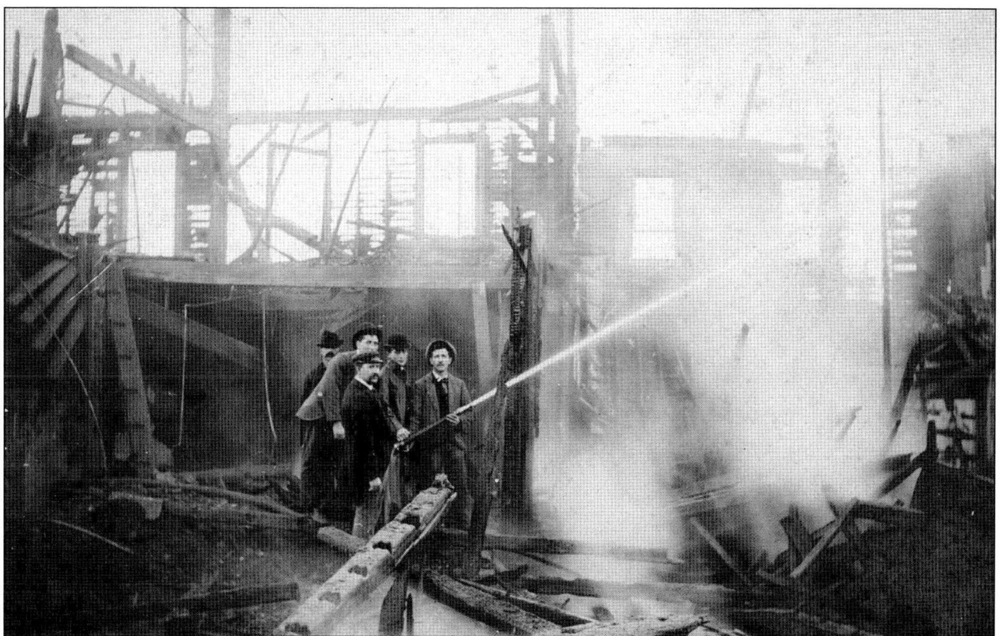

When fire destroyed the Galt House in 1899, it had fallen into such disrepair that few citizens mourned its loss. The fire brought out every firefighter and every firefighting apparatus in the city, but to no avail. One newspaper account notes, "There was a terrible loss of life—cockroaches, bedbugs and rodents by the million were burned to death."

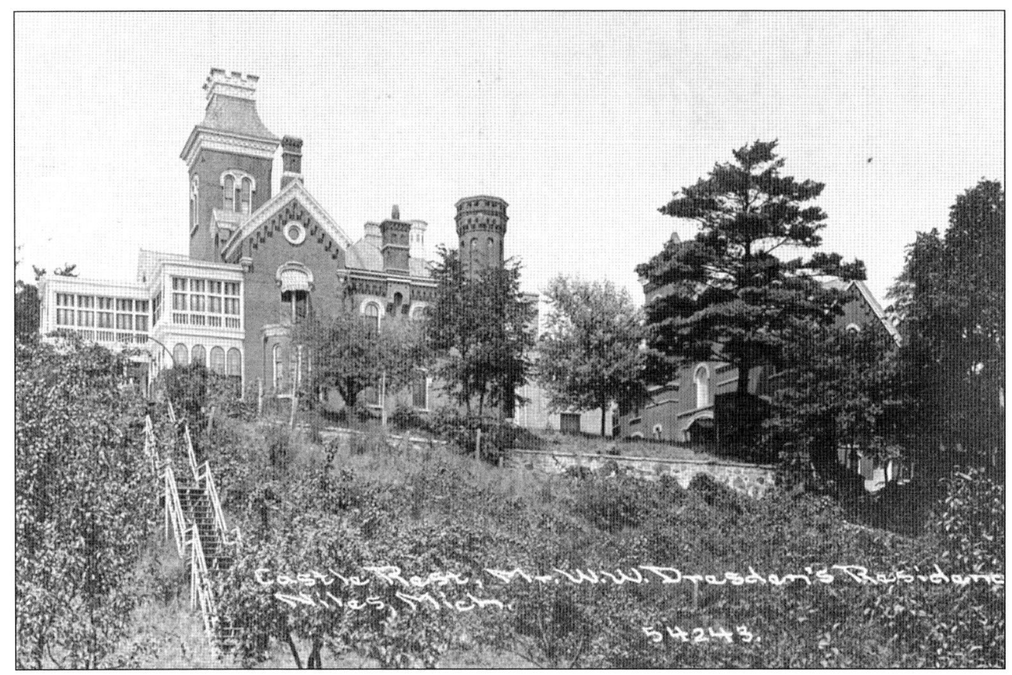

G.A. Colby built his home on the west bank of the St. Joseph River during the 1860s. William Wallace Dresden later purchased and remodeled the home, naming it Castle Rest. In 1925, Francis and Jennie Plym bought the house so that it could be converted into a hospital. The Niles community raised funds to create Pawating Hospital.

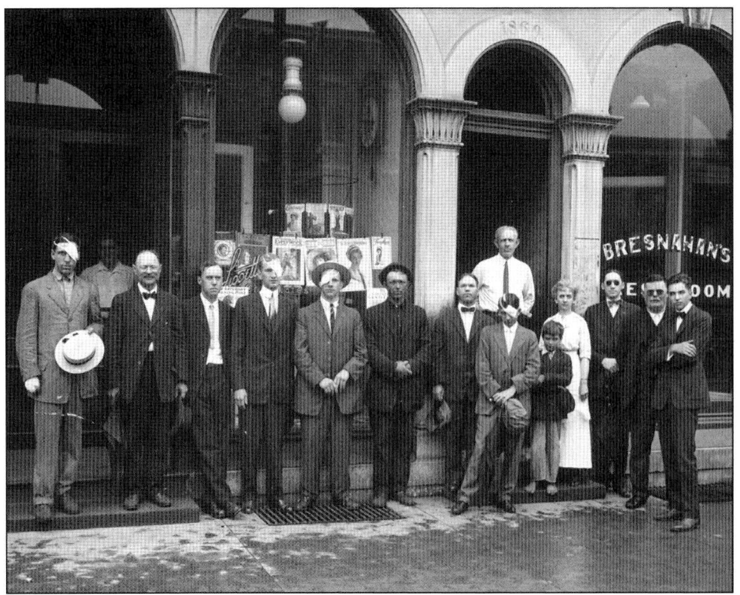

Dr. Frederick Nathaniel Bonine operated an eye clinic in Niles for almost 40 years. Thousands of patients traveled long distances to his clinic, looking for a miracle cure for cataracts. Reportedly, he never turned away patients who could not pay. Here he is pictured outside his clinic in a white shirt and tie. His nurse Grace Lombard stands next to him. Notice the eye bandages on the patients.

Skalla Furniture had a factory on the west side of town. This photograph from 1880 shows their salesroom at the southeast corner of Second and Sycamore Streets. Joseph Skalla started as a cabinet maker, crafting in his own home. He went on to produce everything from fine household furniture to office fittings, which he sold at his downtown shop. His son Paul Skalla served as the city's undertaker for many years.

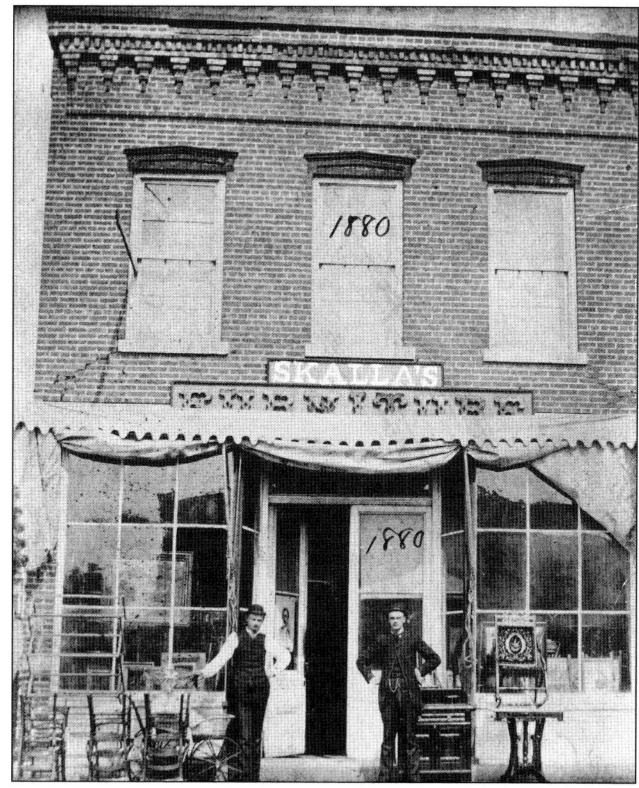

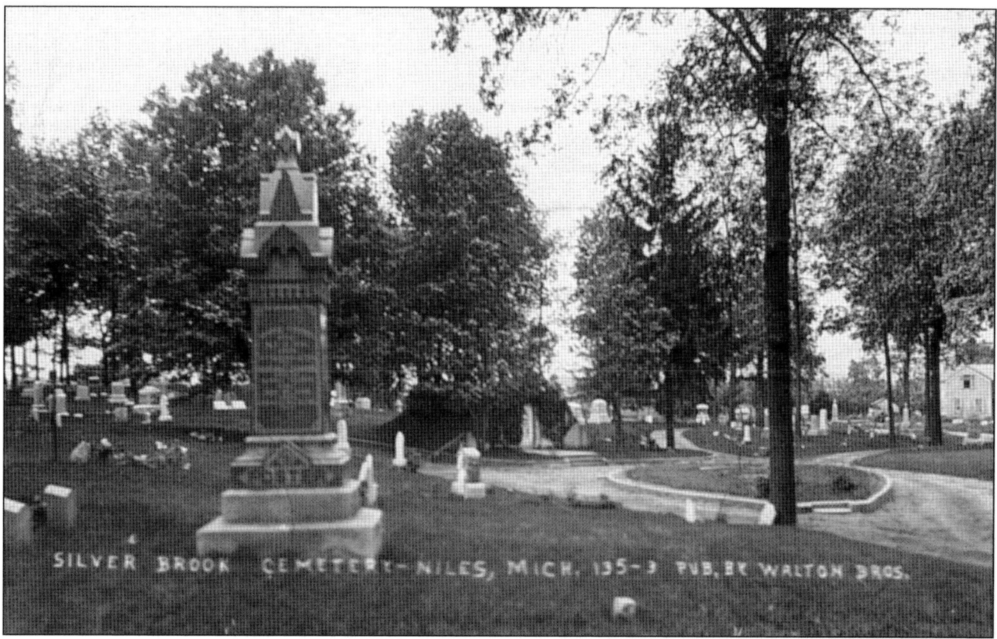

In 1836, George Bond purchased several acres for use as a public cemetery. Named after the brook that runs through the property, Silverbrook Cemetery lots opened to public sale in 1838. Since that time, several additions have been made to the original "Bond" section. This postcard, mailed in 1910, shows a view of the "City" section. The receiving vault, built in 1869, is visible in the background.

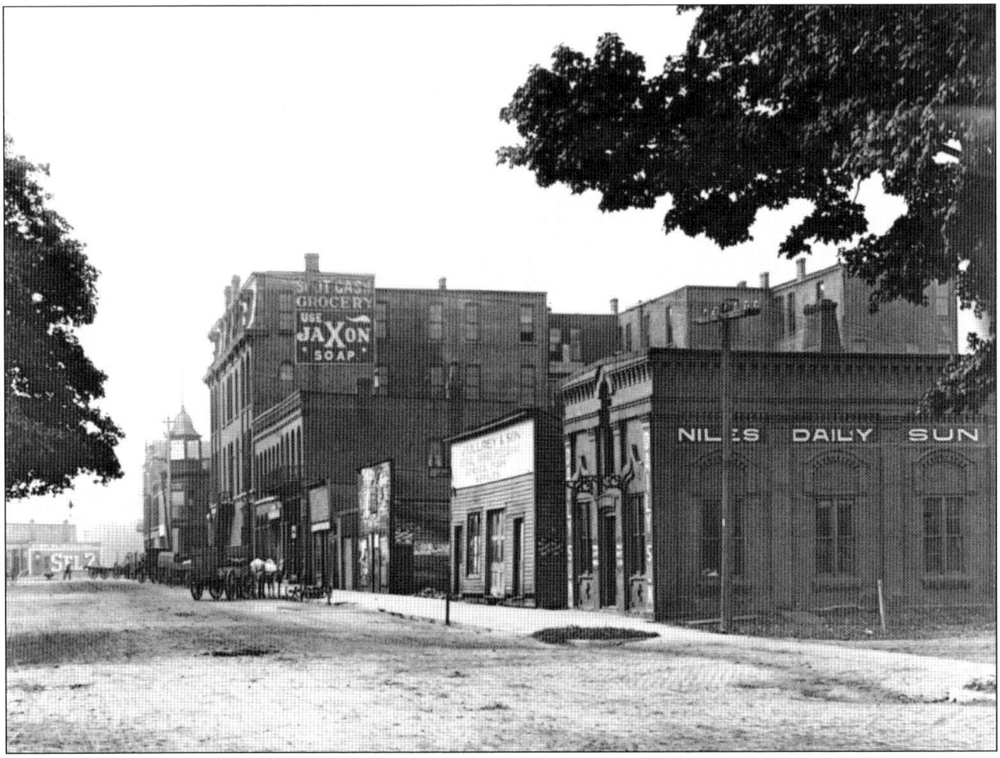

The *Niles Daily Sun* office was located on Front Street, here with a view toward Main Street. The *Sun* succeeded another paper called the *Niles Republican* in 1890 and ran until 1919 when it merged with the *Niles Daily Star*. In the late 1800s, several newspapers were printed locally, including the *Niles Republican*, *Niles Democrat*, *Niles Weekly Mirror*, *Niles Daily Sun*, and *Niles Daily Star*.

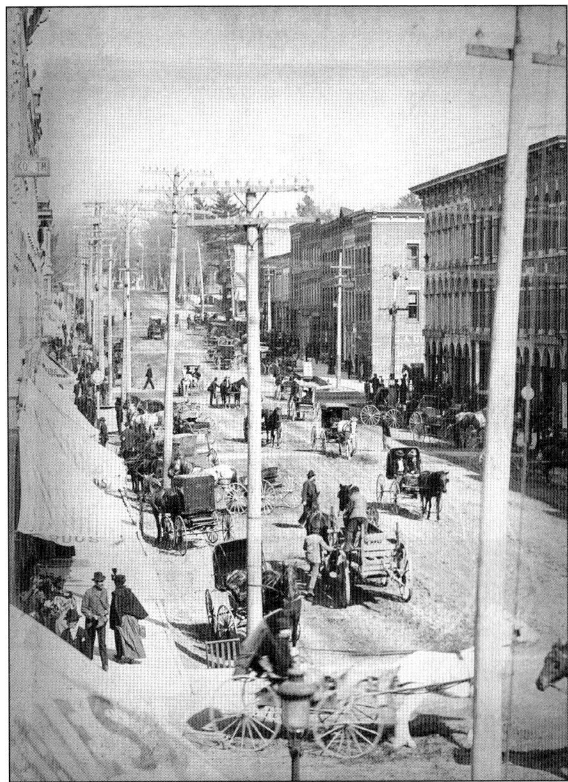

This c. 1899 scene shows a busy Main Street looking east from Front Street at a time when people traveled by horse and buggy. Hitching posts lined the road for parking. Progress is happening, however—note the electric poles. Electricity came to Niles in the year 1890 and powered the downtown.

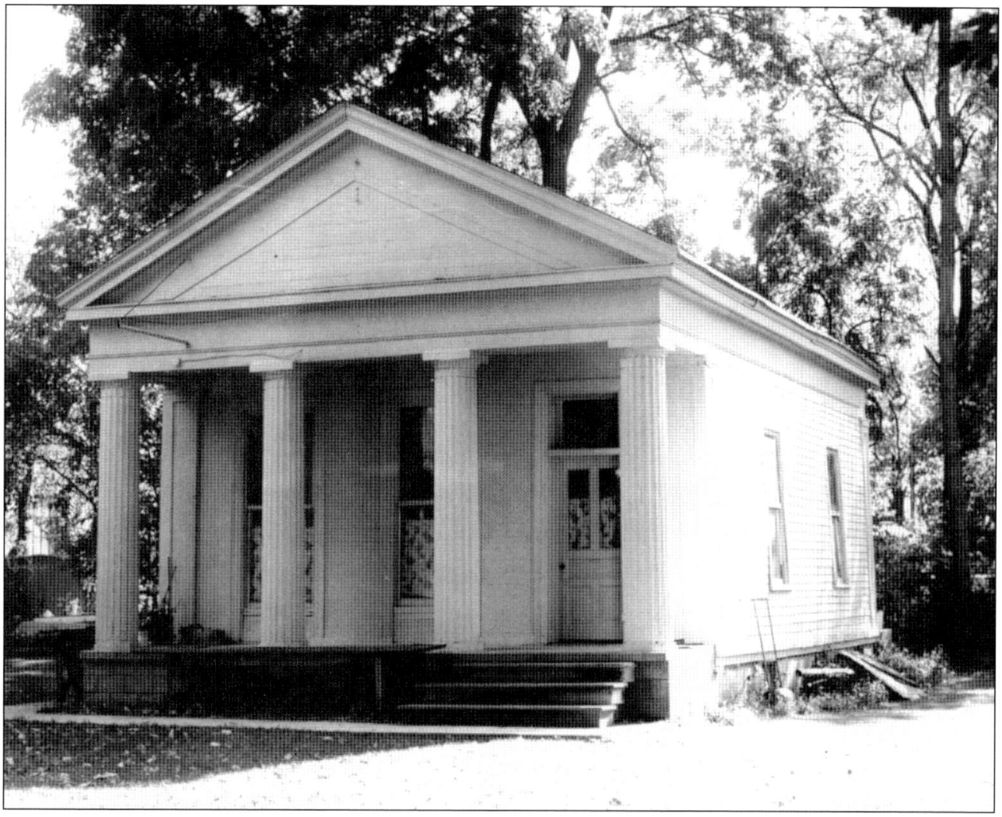

Built in 1842 for Rodney C. Paine, this Greek Revival building originally sat at Third and Main Streets. An agency of Detroit's Farmers and Mechanics Bank until 1852, afterward it operated privately as Paine's Bank. Paine served as Niles Village president, mayor, state senator, and on the Board of Public Schools. In 1973, Paine's Bank was listed in the National Register of Historic Places.

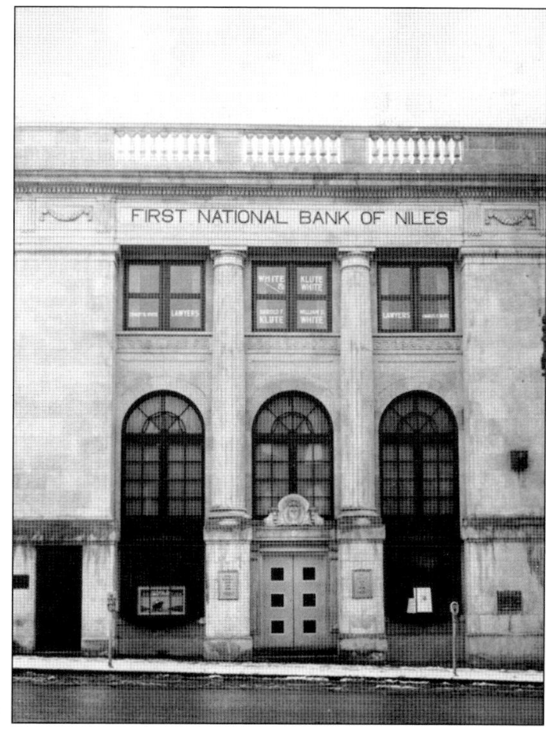

Located at 210 East Main Street, the First National Bank of Niles was organized in 1933 with 10 employees. The bank originally operated out of this sole location, but as the business expanded, new branches opened throughout Berrien and Cass Counties. In 1962, this building was demolished, making way for a modern structure for the bank's main office.

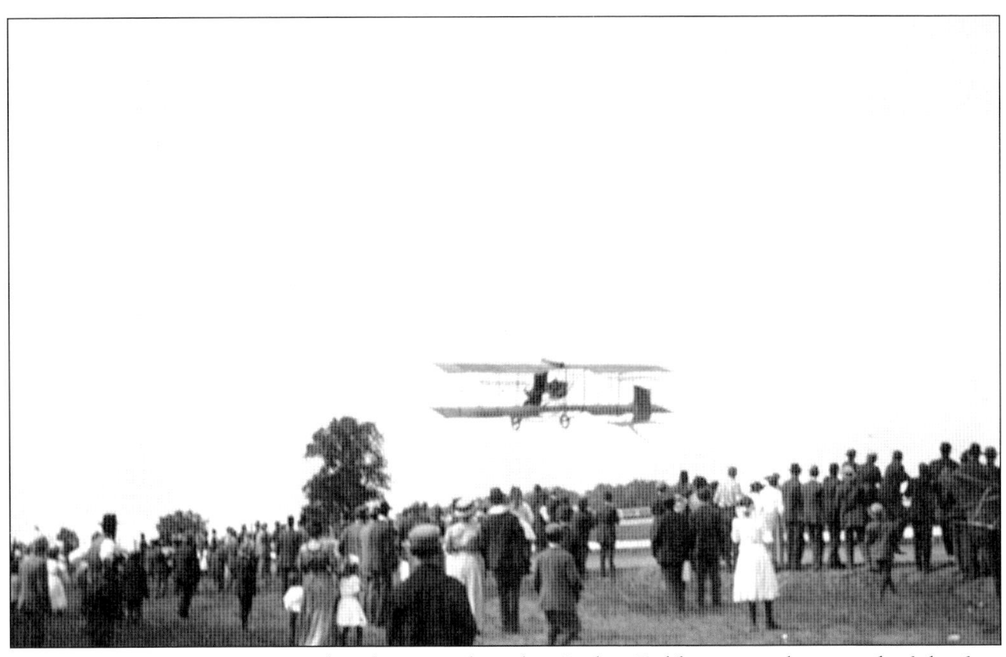

On August 21, 1919, an excited audience gathered at Parker Field to greet the arrival of the first "aeroplane express" to Niles. The biplane delivered a shipment of clothing items from Chicago for the Charles Julius Company. In 1928, the city leased Parker Field for use as a municipal airport. The airport was later expanded and renamed to honor industrialist Jerry Tyler.

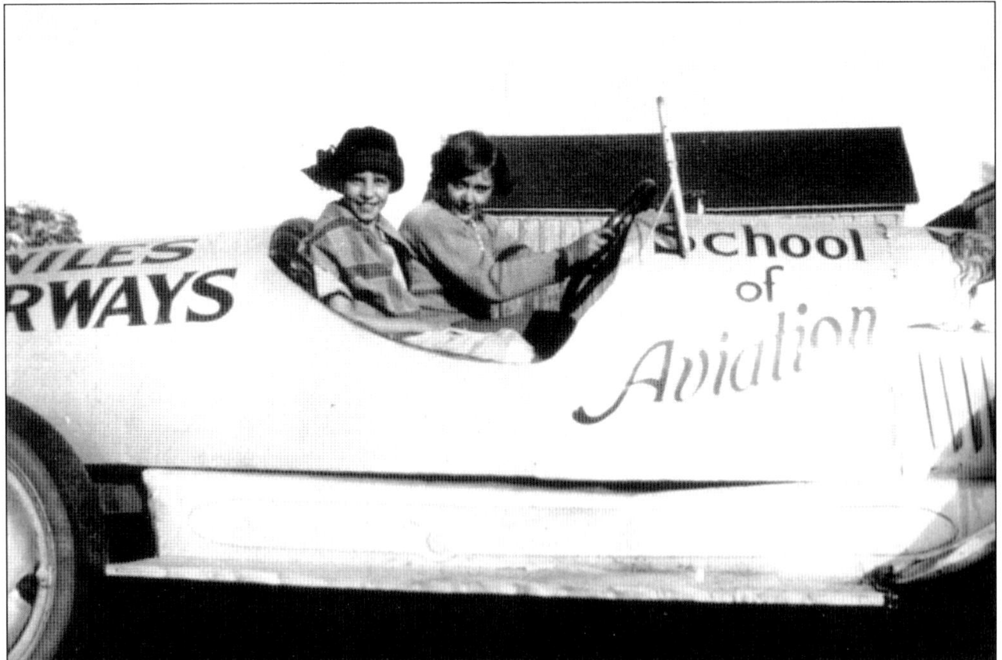

Two women sit in a Niles Airways School of Aviation plane. In 1925, local citizens organized the Niles Airway Company, which offered passenger service to Chicago and flying lessons to those brave enough to attempt this newfangled form of transportation. To attract patrons and prove that air travel was safe, the Niles Airway Company hosted regular exhibitions and races at the local airport.

Four

Business and Industry Boom

Dry goods stores and grist mills of the 19th century gave way to modern business and industry. To attract entrepreneurs, the Niles Improvement Association promoted the city's assets in 1900: two daily and two weekly newspapers, two telegraph companies, five railroads, quality theaters, an energetic historical society, and the best waterpower in the state. These recruitment efforts brought companies such as Kawneer, Simplicity Pattern, and Tyler Refrigeration to Niles, joining homegrown industries like French Paper Company and National Standard.

By the early 20th century, national department stores lined Niles's Main Street. Residents and visitors could shop at J.C. Penney, Montgomery Ward, Woolworth, or a number of smaller specialty shops. Restaurants ranged from soda fountain lunch counters to supper clubs and fine dining establishments. The Ready Theater offered plays, musicals, and the latest films, while the world-class Four Flags Hotel and other inns provided accommodations and hosted community events. An airport was built on the east side of the city.

As the city expanded, shopping centers and strip malls appeared on the south and east edges of Niles. Like many urban areas, downtown struggled to compete, limited by parking and retail space. In an attempt to modernize, buildings were faced with Kawneer metal fronts. Yet businesses continued to spread along the 11th Street corridor toward South Bend. Railroad traffic slowed, and bus services began transporting people. Downtown Niles saw a decline as larger stores moved away. Like much of the nation, the community faced challenges when industries left the area.

In recent years, however, downtown Niles has experienced a resurgence. Much of its 19th-century architecture was restored when metal facades were removed. Buildings now house small shops, cafés, and offices. While industry giants like Kawneer have departed, they left a lasting legacy through philanthropy that helped establish Plym Park, the Niles District Library, and other cultural institutions. Former riverfront warehouses and rail lines have been transformed into recreational areas offering nature trails, playgrounds, and fishing spots. Today, Niles attracts visitors for its history and culture, while residents enjoy its enduring small-town charm and character.

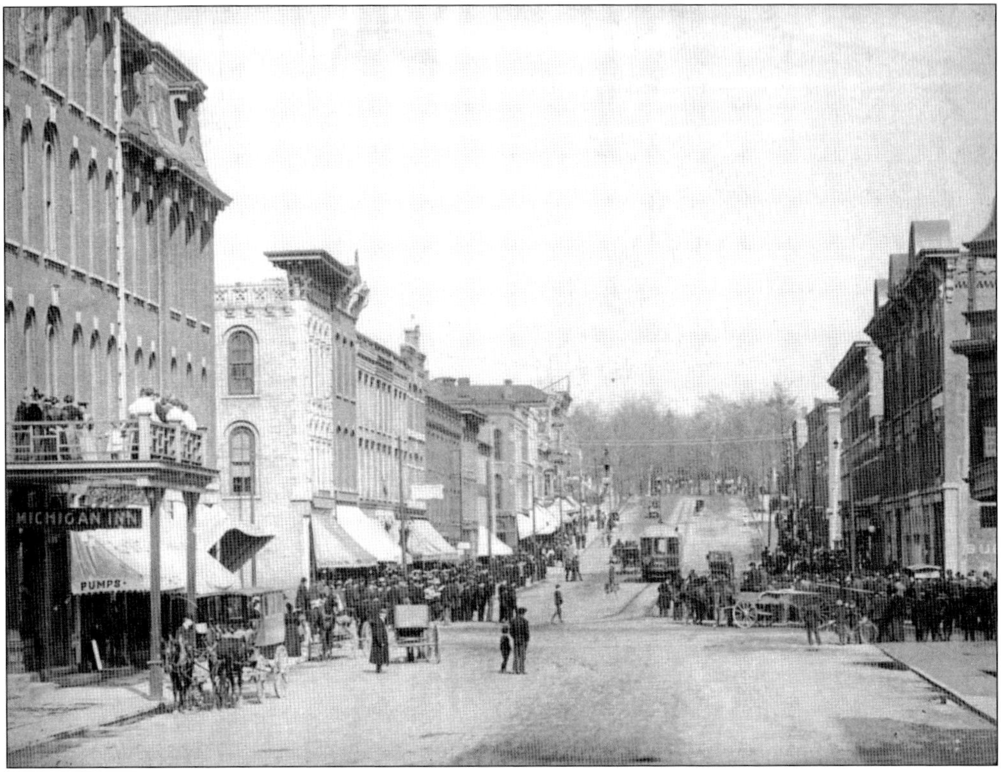

This view looking east on Main Street shows a busy downtown district in 1906. Horse-drawn vehicles are still prominent on Main Street, with the interurban streetcar also transporting passengers through town. A crowd is gathered on the balcony of the Michigan Inn. Originally called the Reading House, the Michigan Inn opened at the corner of Main and Front Streets in 1874.

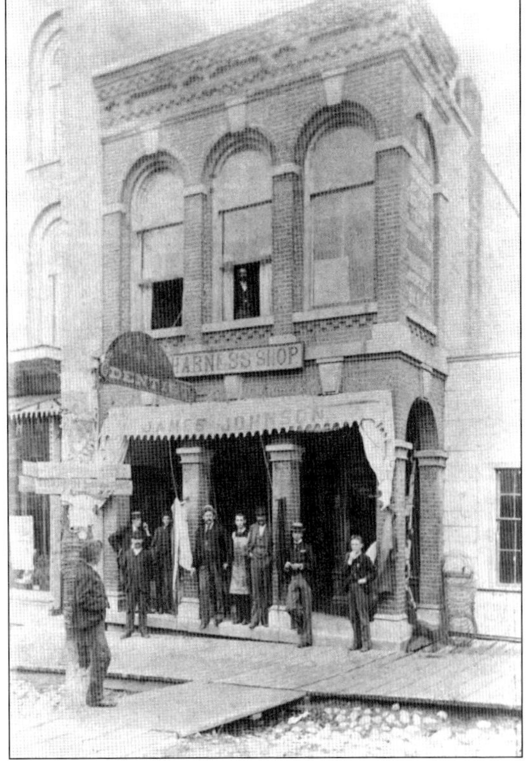

A group stands on the plank sidewalks outside Johnson Harness Shop on Main Street in the 1880s. Born in 1852, James Johnson started his business selling leather goods in 1876, originally partnering with E.E. Garlanger. James Johnson retired in 1930, and his son took over the business, operating it until 1946, when the building was sold to Bill Falvey, who opened a clothing store for men and boys.

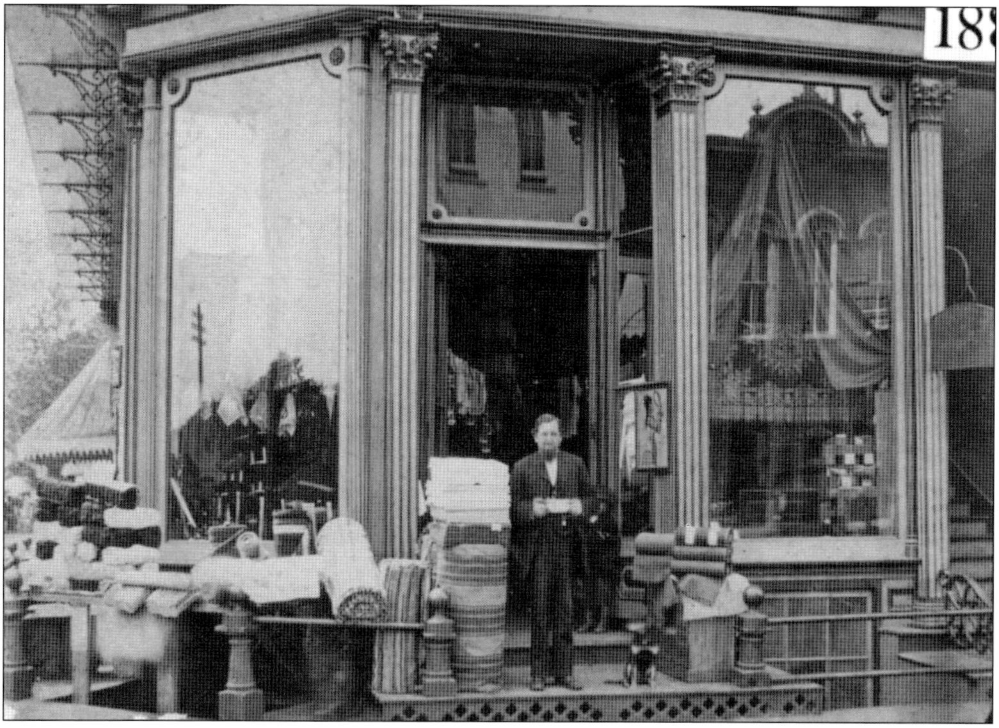

John Henkel opened a store on the 200 block of Main Street with his son Henry in 1883. Ads promised dry goods and fancy notions; everything needed in any family: silks, laces, cloaks, carpets, and more. Finding success in Niles, the Henkels opened a second store in Mishawaka. A newspaper ad noted, "A person who drinks freely of sweet cider should get his suspenders [at] Henkel and Son."

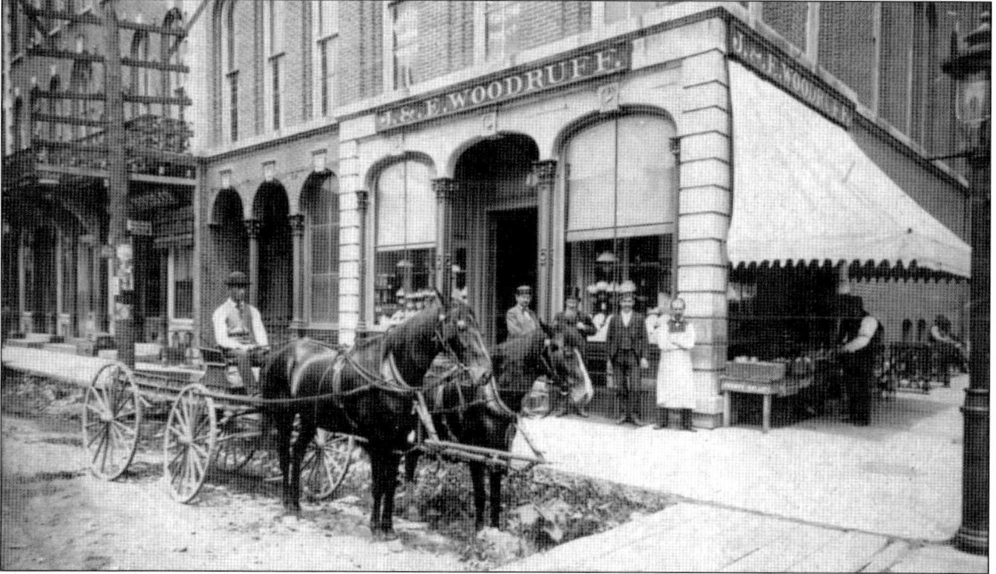

Part of the large Woodruff family, brothers John and Edgar moved to Niles from New York in the early 1850s. They took over a grocery stand at the northwest corner of Main and Front Streets, which was replaced by this four-story brick building in 1868. J&E Woodruff sold a variety of "staple and fancy groceries" and remained in the family until 1896, when it was sold to E.B. Winter.

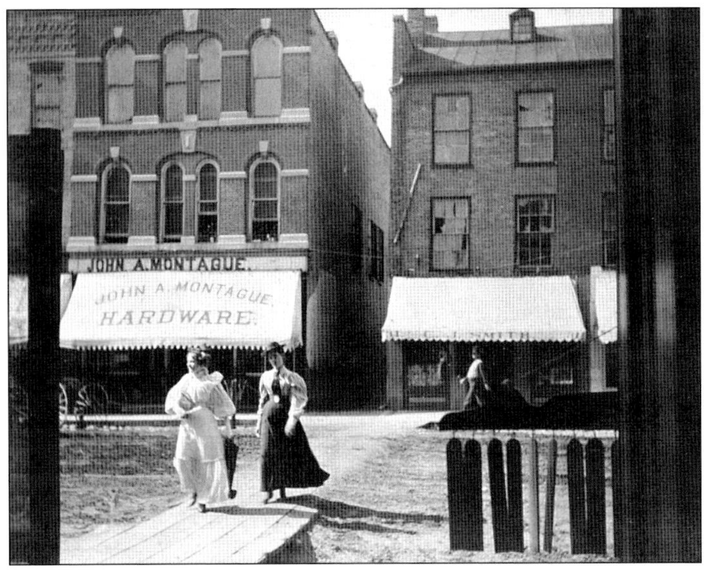

Ruth Reddick and Edith Miller traverse Main Street on the plank crosswalks near Montague Hardware in 1895. John Montague moved to Niles in 1867 and established his hardware store in the 100 block of Main Street. After John's death in 1903, his son Charles took over the business. Montague Hardware closed in 1933. That same year, Andy's Stop and Shop, an office supply and hobby store, moved into the building.

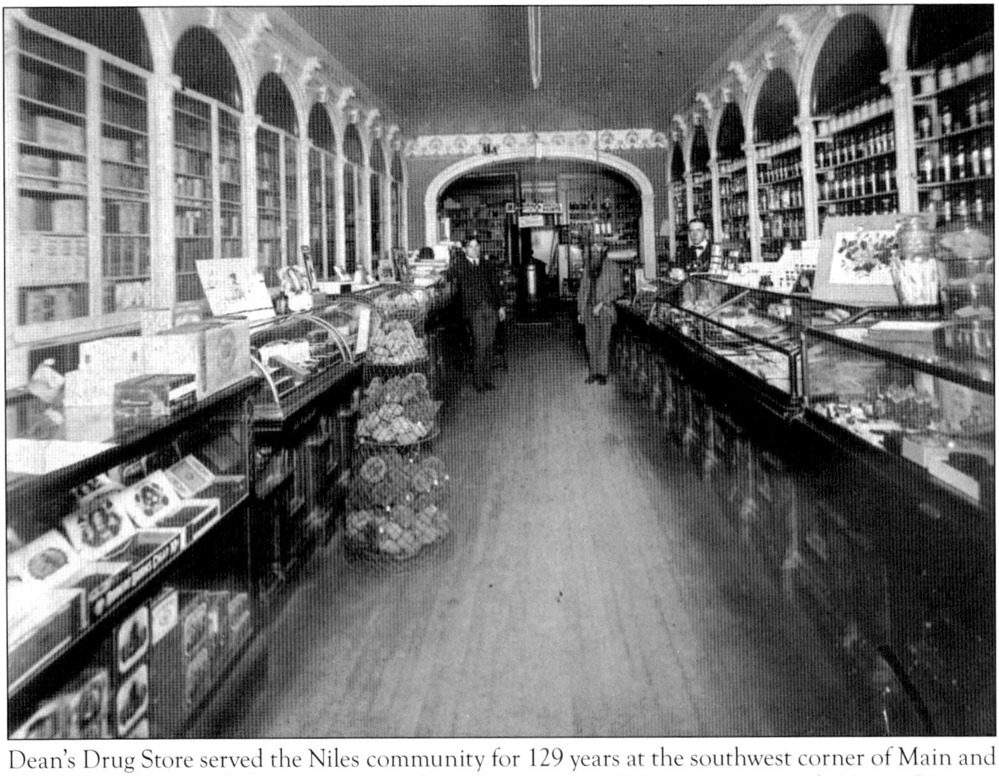

Dean's Drug Store served the Niles community for 129 years at the southwest corner of Main and Second Streets. Joseph Larimore opened the business in 1838, partnering with Henry Dean in 1860. Originally called Larimore's, the name changed to Dean's Drug Store in the 1880s, around the time of this photograph. The business closed in 1967. At one time, Dr. Fred Bonine's eye clinic operated in a space above the store.

Niles businessmen join Albert Green, seated in the chair without a hat, outside his shop on Main Street to read the daily news in 1899. Located below the Niles Opera House, A. Green sold men's and boys' suits, coats, and accessories, with "a guarantee of style, service and satisfaction." The photograph shows a nice selection of the shop's hats. Why is its proprietor hatless?

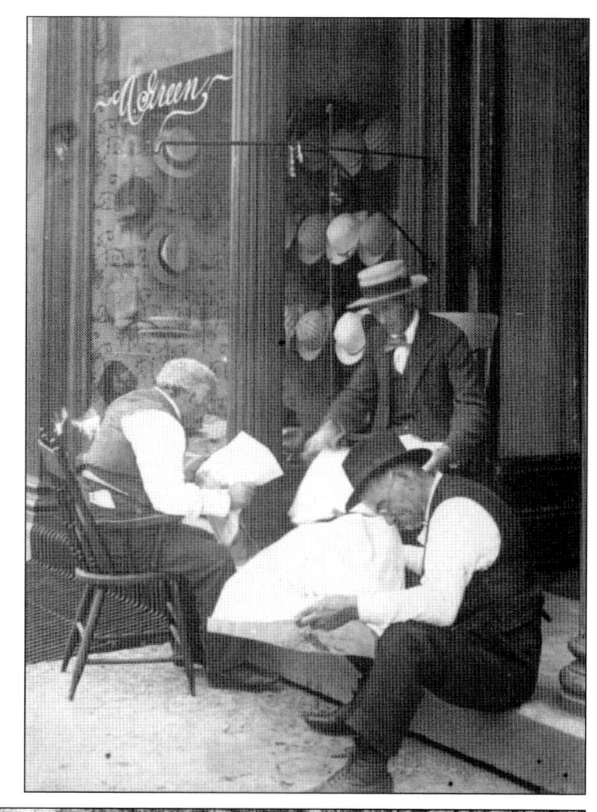

John Weitz sits outside his shoe shop in a small building owned by W.A. Reddick at the northeast end of the Main Street Bridge. In 1913, Weitz took over the Bowerman shoe shop at this location. His business promised "Repairing done while U wait." A fire forced the business to close for a few months in early 1914, but Weitz reopened his shop and ran it until 1916.

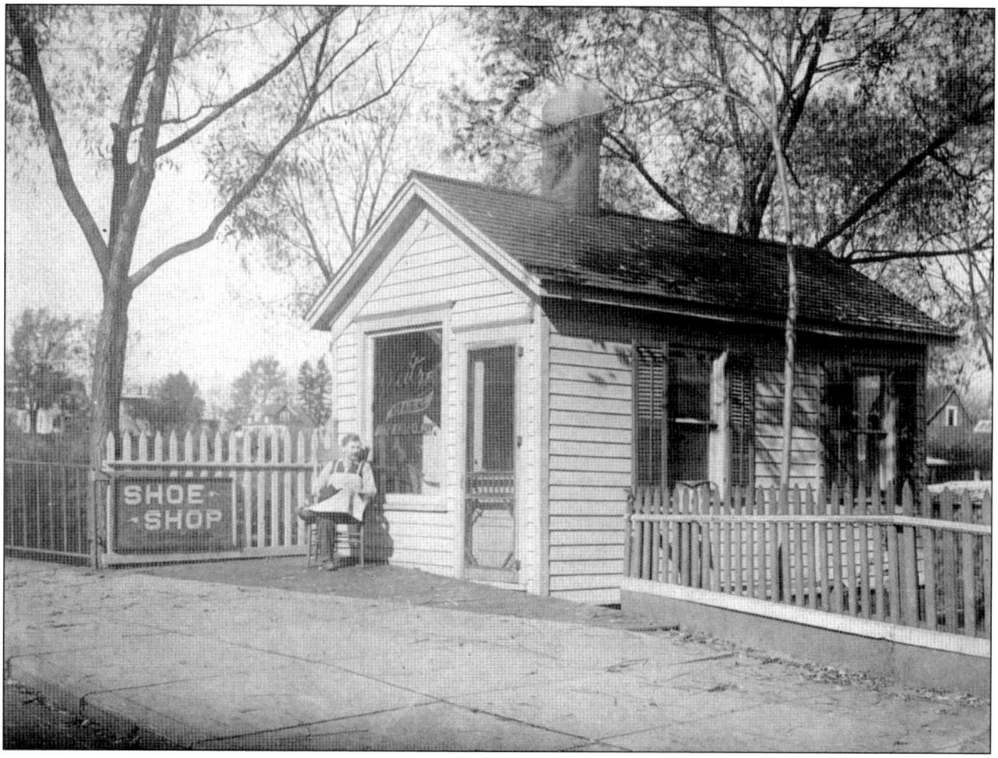

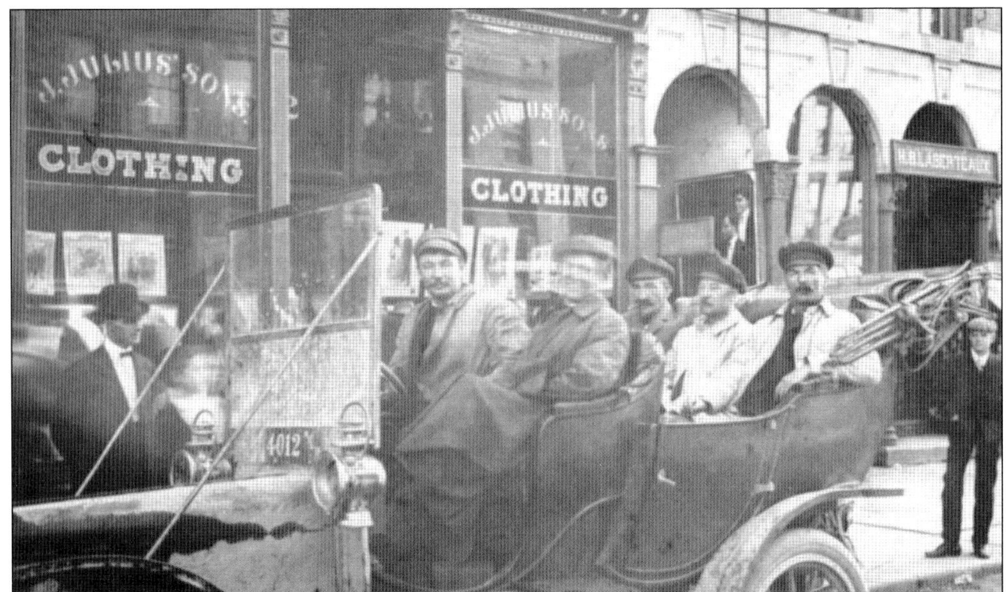

Jim Hatch, Charles Julius, Charles Wood, Bascorn Parker, and an unidentified gentleman park downtown around 1908. Originally founded by Jacob Julius in 1863, his sons Charles and Louis later joined the business and changed the name to J. Julius' Sons. Hatch, the driver, operated the first automobile dealership in Niles and often chauffeured people around.

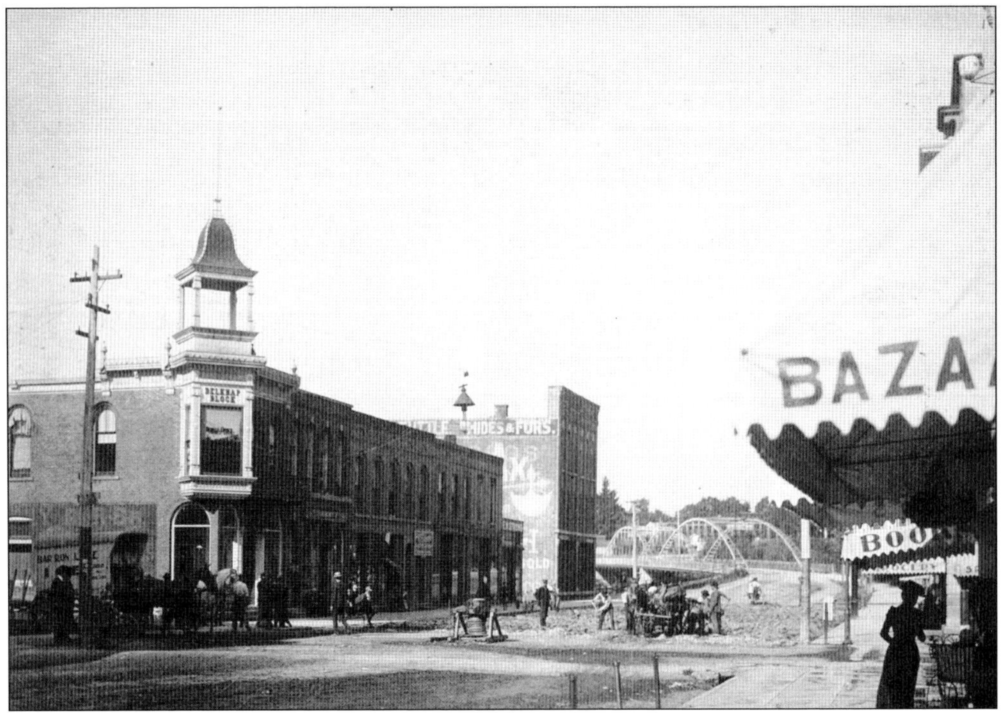

This view looks west on Main Street during the street paving in 1902. The Belknap block, with its grand cupola, was built at the southwest corner of Main and Front Streets in 1894. The block was named for the Belknap family and included offices for Dr. Simeon Belknap and his son, Fred. Another son, also named Simeon, operated a jewelry store in the building.

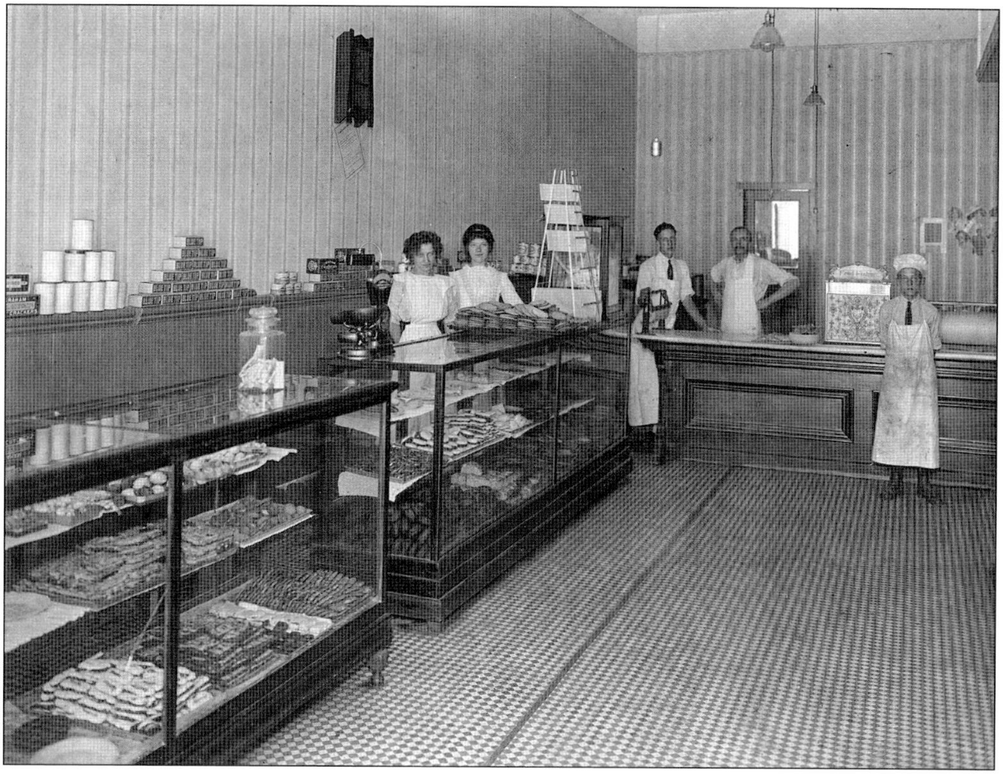

In 1909, Fred Fisher moved his baking business to the Michigan Inn. Fisher's sold cookies, cream puffs, Thanksgiving pies, and his specialty, salt-rising bread, which did not use yeast. By 1915, Fisher closed his bakery in Niles and moved back to South Bend, citing low profits, especially during the winter months.

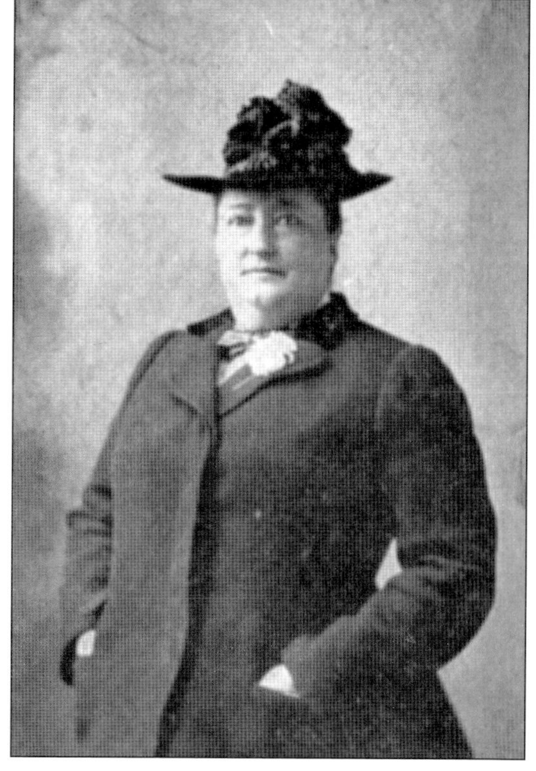

Niles businesswoman Kate Nobles is pictured here. Chewing gum was the most successful product of the Kate W. Nobles Manufacturing Company, located at Fourth and Main Streets during the 1890s. Not only did women lead this burgeoning business, but they also employed women almost exclusively in their factory. Each gum box proudly noted, "Originated and Manufactured by Women."

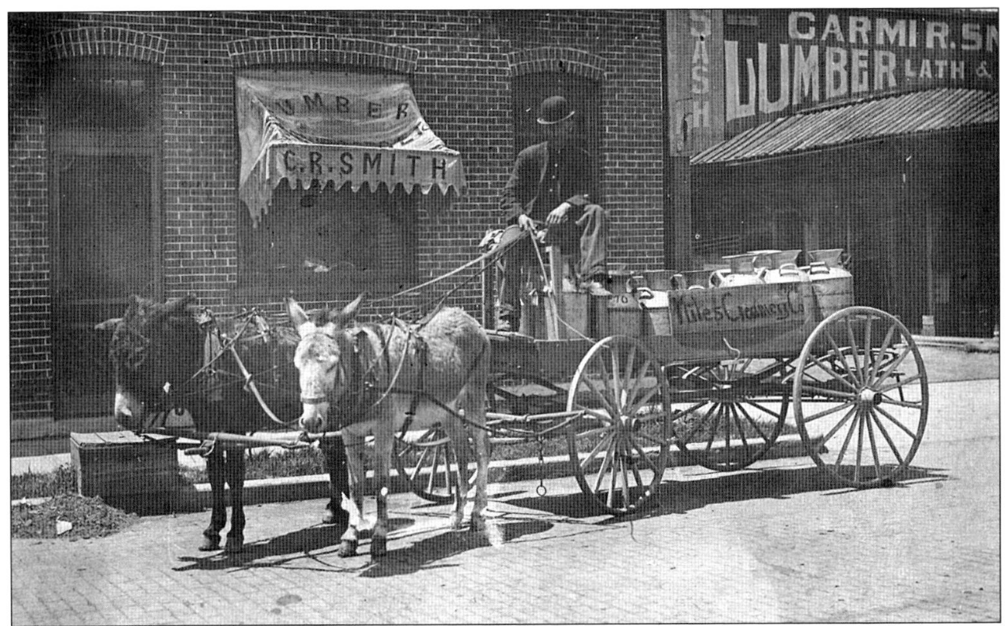

A Niles Creamery wagon waits in front of the Carmi Smith Lumber Yard in the early 1900s. Smith operated a lumber yard on Front Street and also managed the Niles Creamery. Smith helped organize the Niles Hotel Company to raise money for building the Four Flags Hotel. He served as Niles's mayor and was active with the Niles Business Men's Association, a group that urged industries, including Kawneer, to relocate to the city.

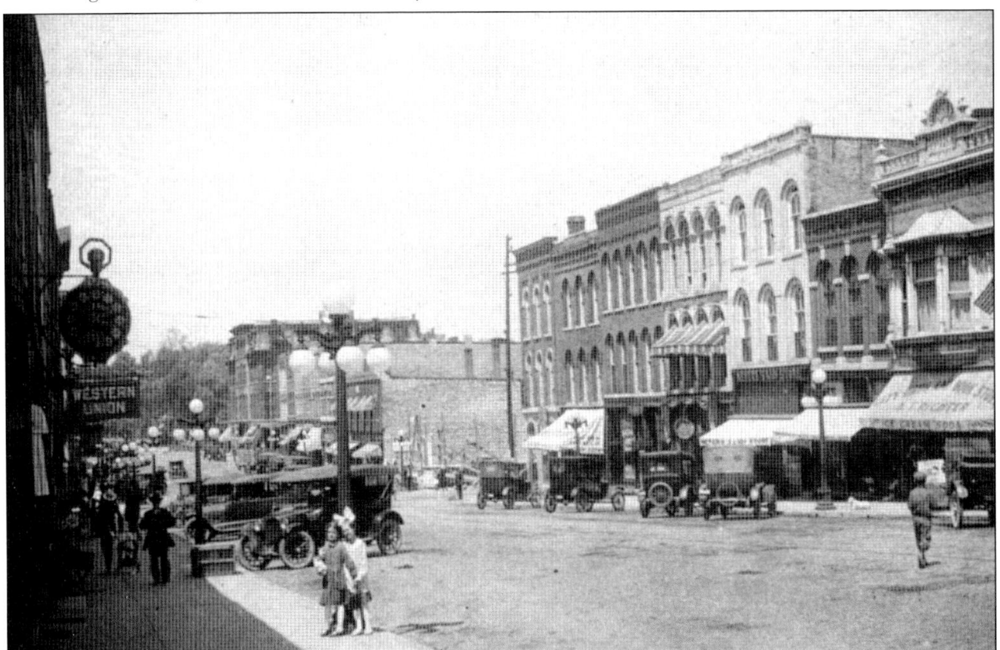

Two girls pose for a picture on a busy Main Street sidewalk in 1919. The image was taken near Third Street and shows a great view of the businesses looking west toward the St. Joseph River. In the background, a new building to house Newman and Snell Bank at the northwest corner of Main and Second Streets is under construction.

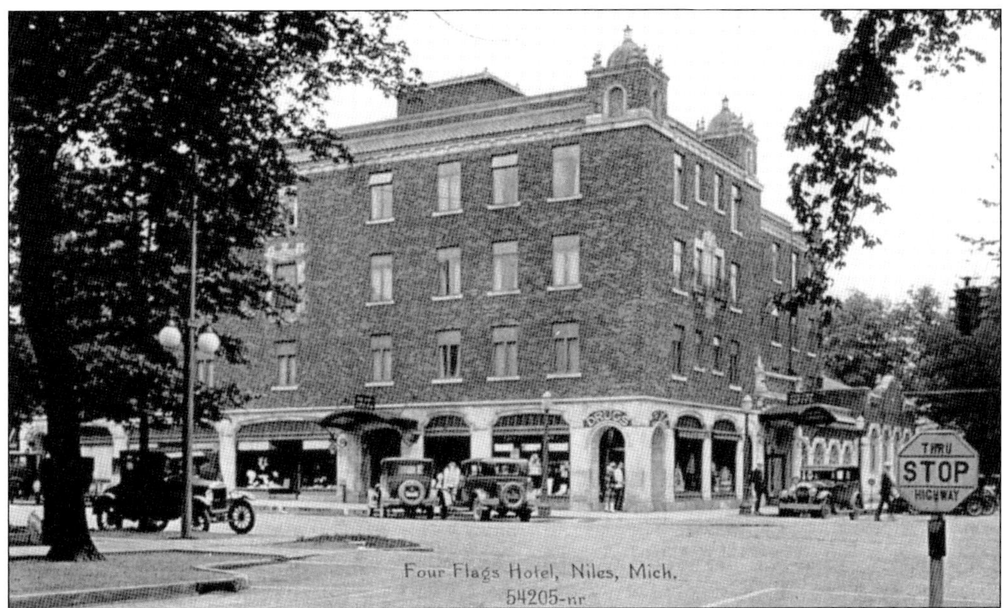

The Four Flags Hotel opened at the site of the former Pike House in 1926. To fund this endeavor, the Niles Hotel Company, a coalition consisting of many local industry and civic leaders, raised $350,000 in subscriptions from the community. The premier hotel in Niles, the Four Flags included a modern elevator, a switchboard with a direct line to the train depot, and the beautiful Queen Anne Room for large gatherings.

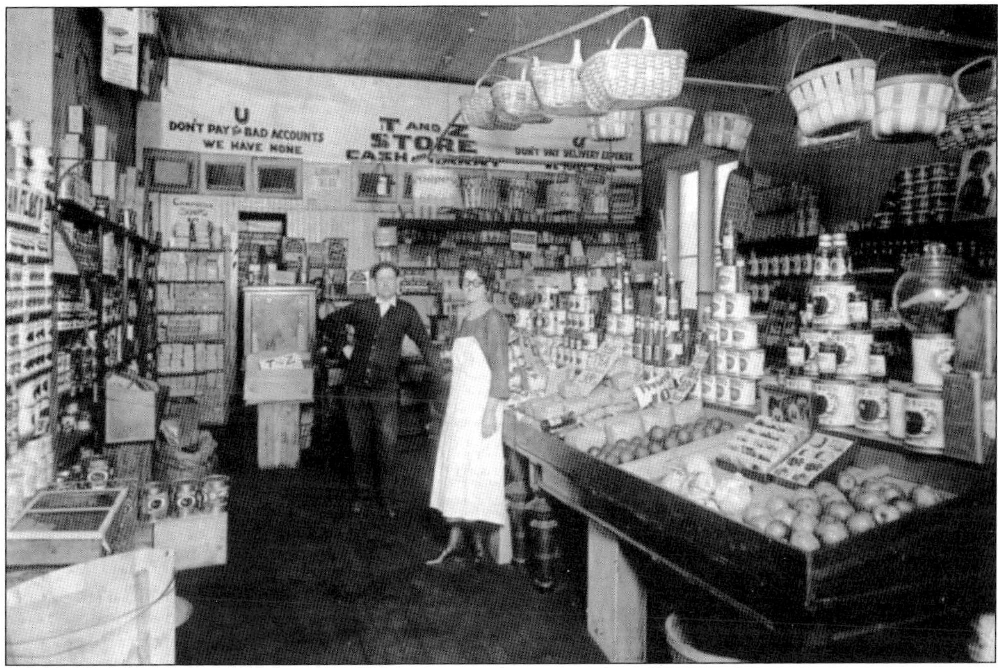

In the early 1920s, the T&Z stores opened on Second Street and at Ninth and Main Streets. Named after the proprietors, Charles Tuttle and Charles H. Zwergel, the store promoted itself as a "bargain cash and carry business," selling a variety of groceries, including fresh fruit, candy, and Campbell's Soup products.

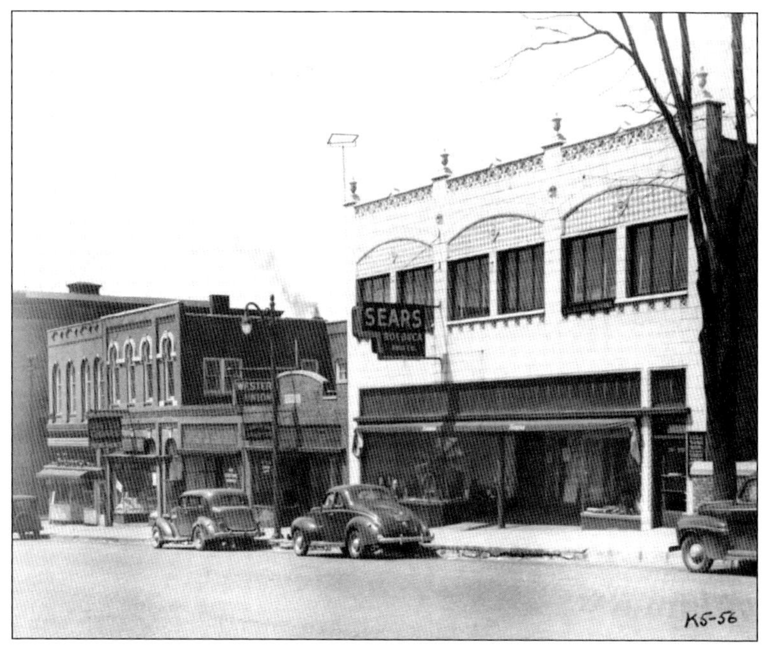

From 1940 to 1945, Sears, Roebuck and Company occupied this building at 311 E. Main Street, between Third and Fourth Streets. The business closed in 1945, but the company reopened in a new location at the corner of Fifth and Sycamore in 1950. The business left Niles in early 1979. This photograph also shows the other shops on the block, including Blackmond's Jewelry.

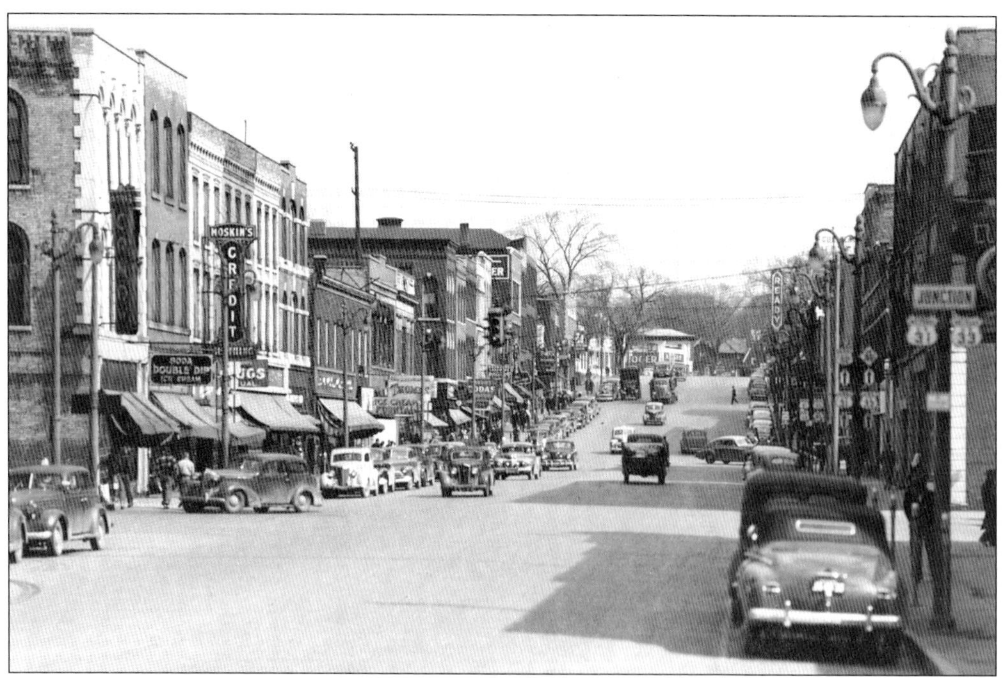

In 1910, Paris Candy Works opened at the northeast corner of Main and Front Streets. Operated by James Patterson, the store sold homemade candy and other items at its soda fountain. In the late 1920s, Patterson purchased another storefront at 220 East Main Street. In the early 1950s, the storefront at 101 E. Main Street closed, but business continued at the second location.

In 1919, Frank Marazita purchased the building on the southwest corner of Main and Third Streets, where his family ran the Niles Fruit Company until 1939. That year, the store was renovated to house Veni's Sweet Shop, operated by Frank's son Joseph and his wife, Josephine. Still a familiar name in Niles, Veni's used to serve up lunch and soda in addition to a variety of treats.

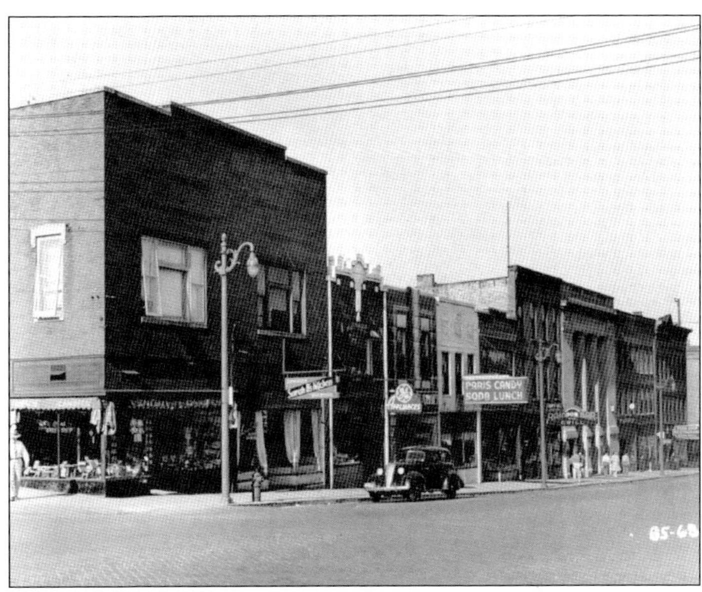

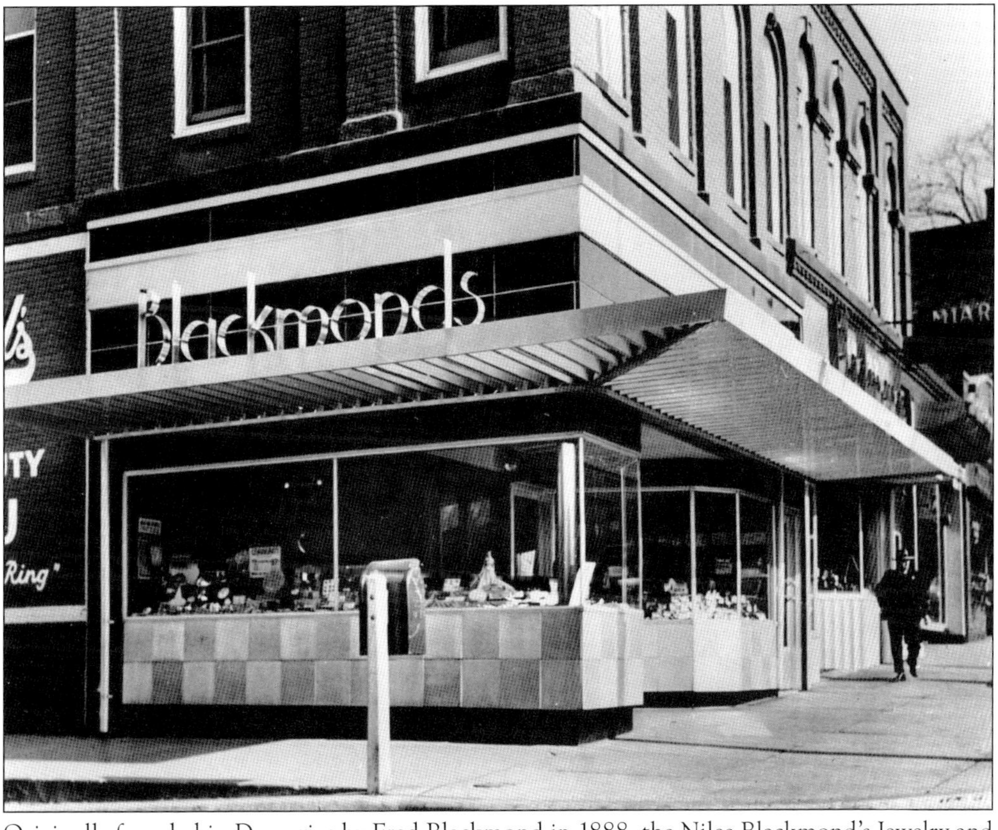

Originally founded in Dowagiac by Fred Blackmond in 1888, the Niles Blackmond's Jewelry and Optical Store opened in 1921. In 1941, the business moved to this location on the northwest corner of Main and Third Streets. Blackmond's new store boasted a state-of-the-art Kawneer front. The black porcelain, stainless steel, and aluminum features were designed to stand out as a shining example to Kawneer dealers visiting the company's headquarters.

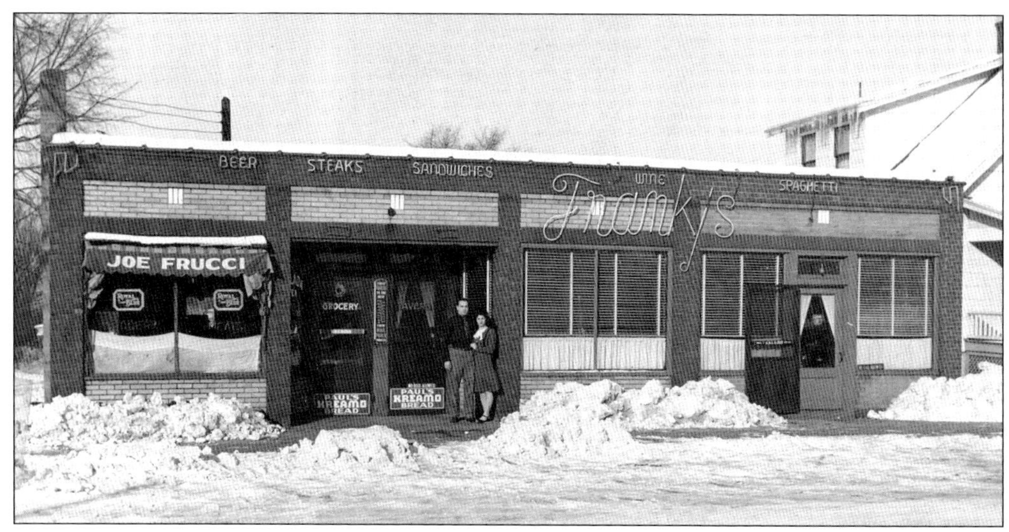

In 1946, Franky Frucci Jr. returned to Niles after serving in World War II and took over his father's grocery and tavern business, changing the name from Lake Street Inn to Franky's. Franky's restaurant was a Niles institution known for its ham sandwiches and Italian-American cuisine. The Frucci family lived in the "Dickereel" neighborhood, an inclusive area on the city's northeast side comprised of German, Italian, and African American families.

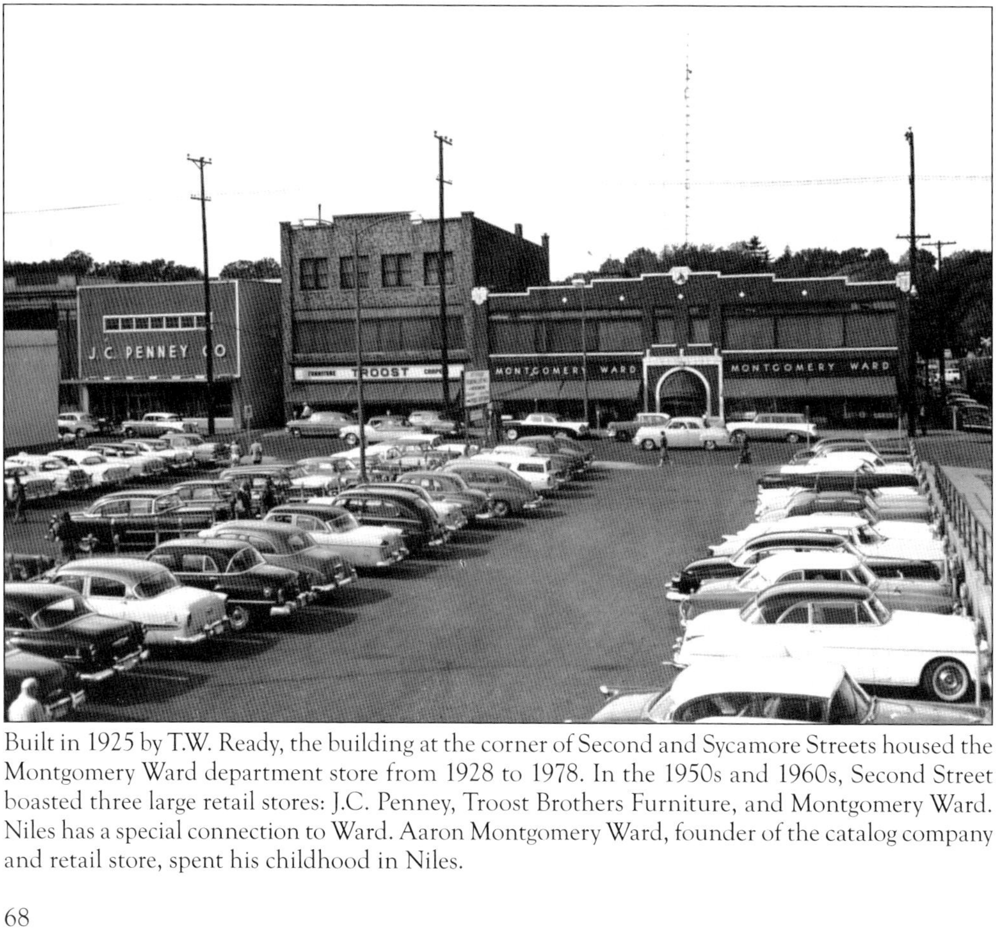

Built in 1925 by T.W. Ready, the building at the corner of Second and Sycamore Streets housed the Montgomery Ward department store from 1928 to 1978. In the 1950s and 1960s, Second Street boasted three large retail stores: J.C. Penney, Troost Brothers Furniture, and Montgomery Ward. Niles has a special connection to Ward. Aaron Montgomery Ward, founder of the catalog company and retail store, spent his childhood in Niles.

Acklee King opened Buster's Riverview Inn on Front Street in 1939. Buster's served a mostly Black clientele and welcomed people who were often turned away from other businesses in the community. In 1971, the State of Michigan recognized Buster's as the oldest tavern to hold a liquor license at the same location in the state. Shortly after this recognition, the building was slated for demolition as part of urban renewal.

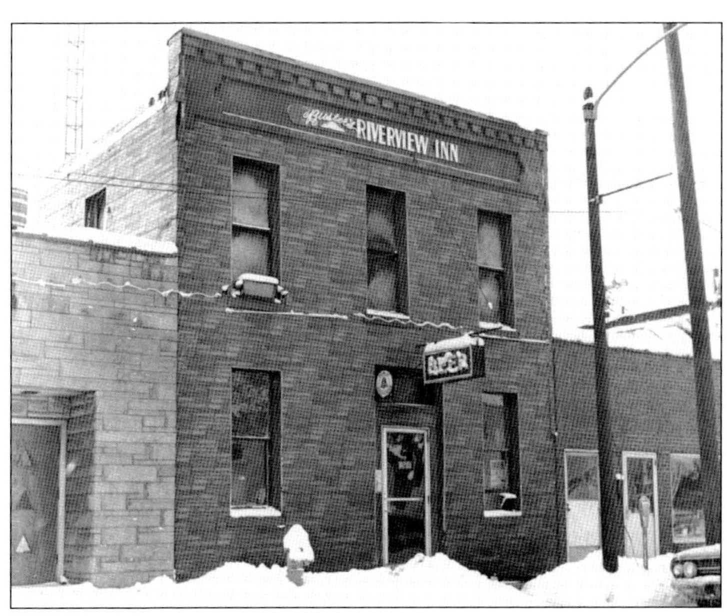

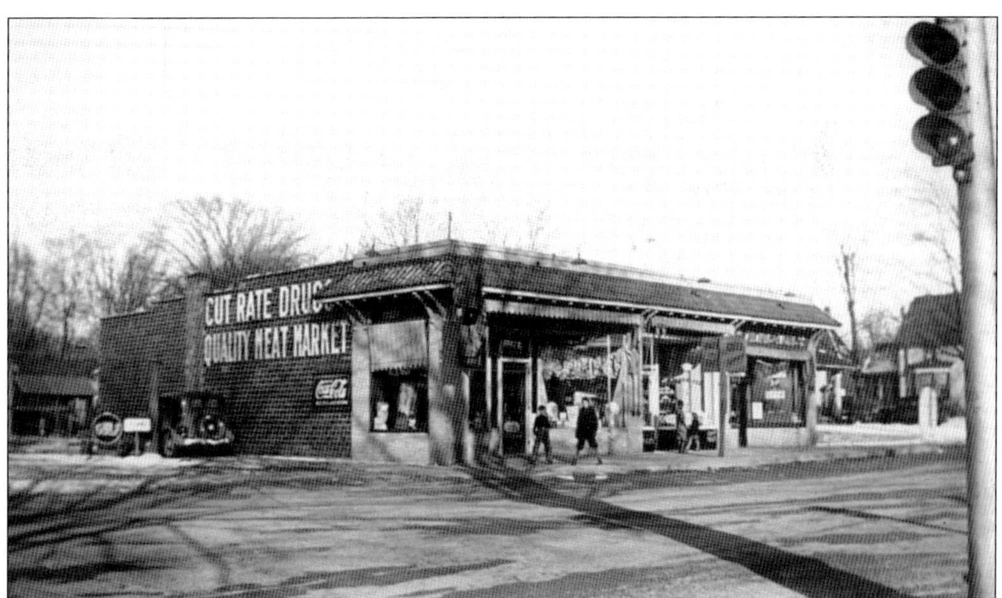

Francis Drolet opened Drolet's Drug Store in the 1920s at Twelfth and Main Streets. The side of the building advertises Cut Rate Drugs and a Quality Meat Market. Drolet's also featured a soda fountain and lunch counter. The package liquor department opened in 1933, the first in Niles after prohibition. This photograph dates to the 1930s. Future proprietor Louis Drolet is seen in the front on the left with one of his brothers.

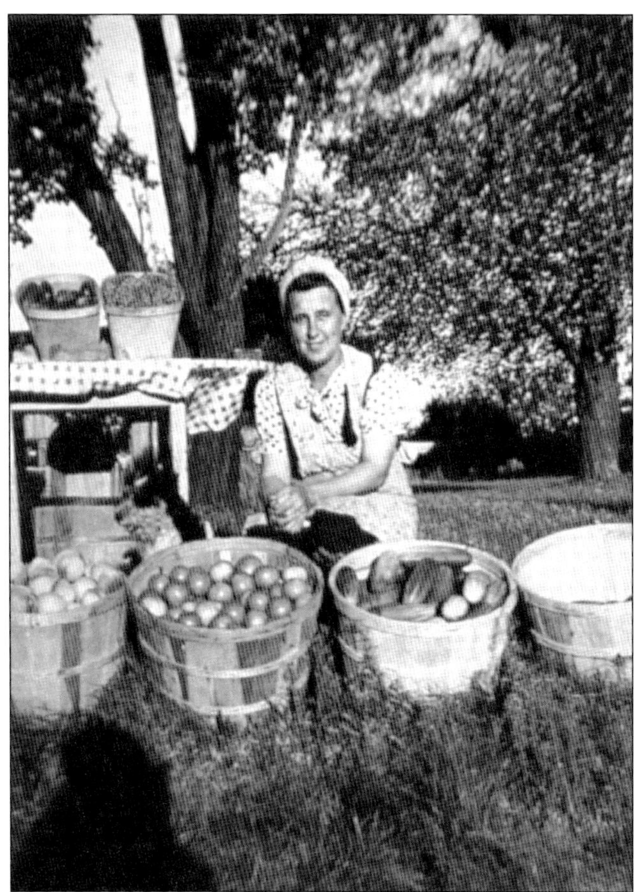

In 1947, Ethan and Rose Shelton founded Shelton Farms. The business stayed in the Shelton family for 77 years as it grew from a roadside stand into a grocery, meat market, and garden store. Shelton's Farm Market closed its doors in December 2024. This photograph shows Rose Shelton selling produce at Shelton's roadside market in the 1940s.

Garden City Fan Company, manufacturers of heating and ventilating products for commercial, industrial, and residential use, originated in Chicago in 1879. In the early 1900s, local city leaders offered industries incentives to relocate to Niles. In 1902, Garden City Fan opened a factory at the corner of Eighth and Wayne Streets near the railroad tracks. The successful industry operated in Niles until 2004.

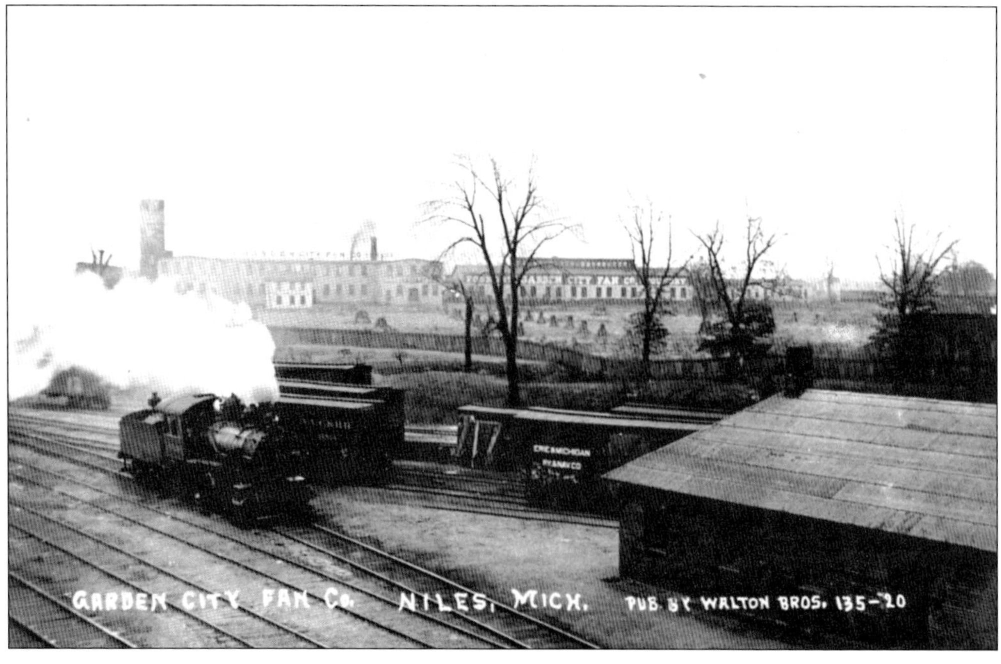

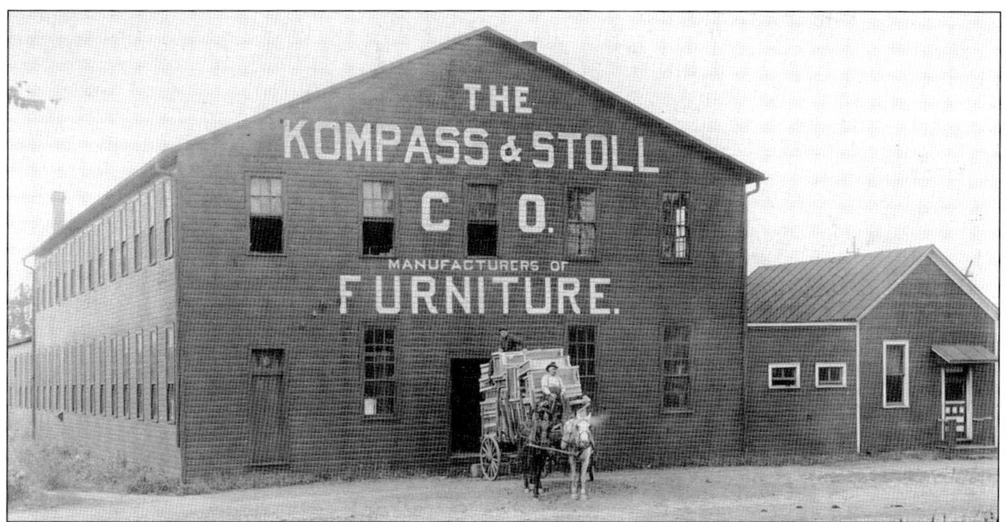

Rudolph Kompass and Matthew Stoll founded their furniture business in Buchanan in 1890. They moved to Niles in 1895, opening a large factory on Pokagon Street near the banks of the St. Joseph River. They had a sample room in Chicago to showcase their products. The industry founders were also active in city government, serving on various boards and committees. Matthew Stoll served three terms as mayor of Niles.

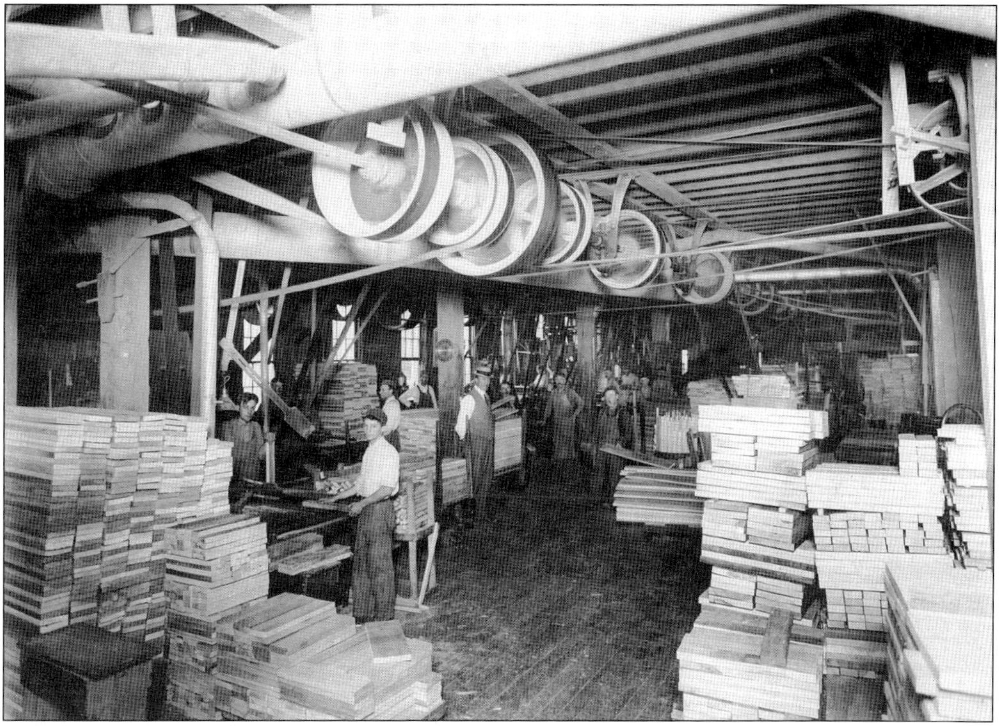

This photograph shows the interior of the Kompass & Stoll factory. The company manufactured furniture, starting with parlor tables and later kitchen pieces. Their Sno-White products, made with a special enamel, earned the firm national recognition. However, the Great Depression led to Kompass and Stoll's receivership in the early 1930s. In 1936, Kawneer Company purchased the shuttered factory.

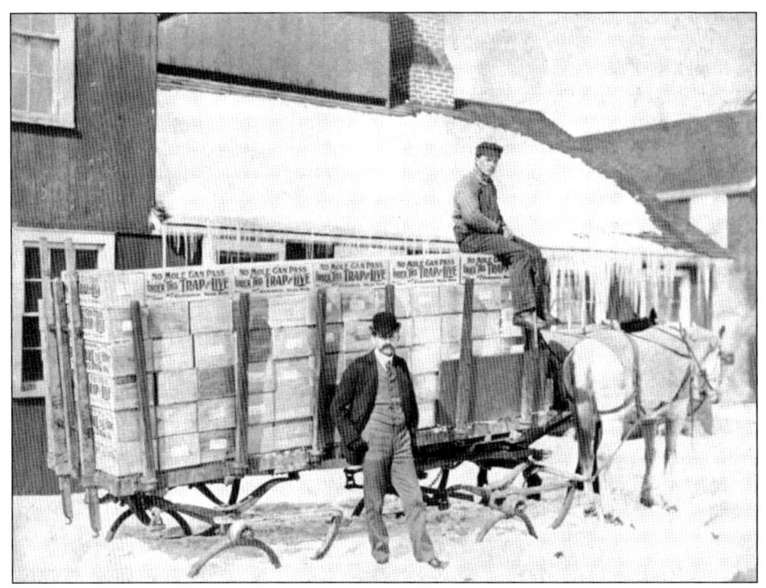

Simeon Belknap stands in front of a wagon loaded with Reddick mole traps, ready to ship. In 1878, William A. Reddick, a charismatic local character, started a wire goods business. The W.A. Reddick Wire and Shovel Factory produced a variety of products, including shovels, mole traps, potato forks, and even doll beds.

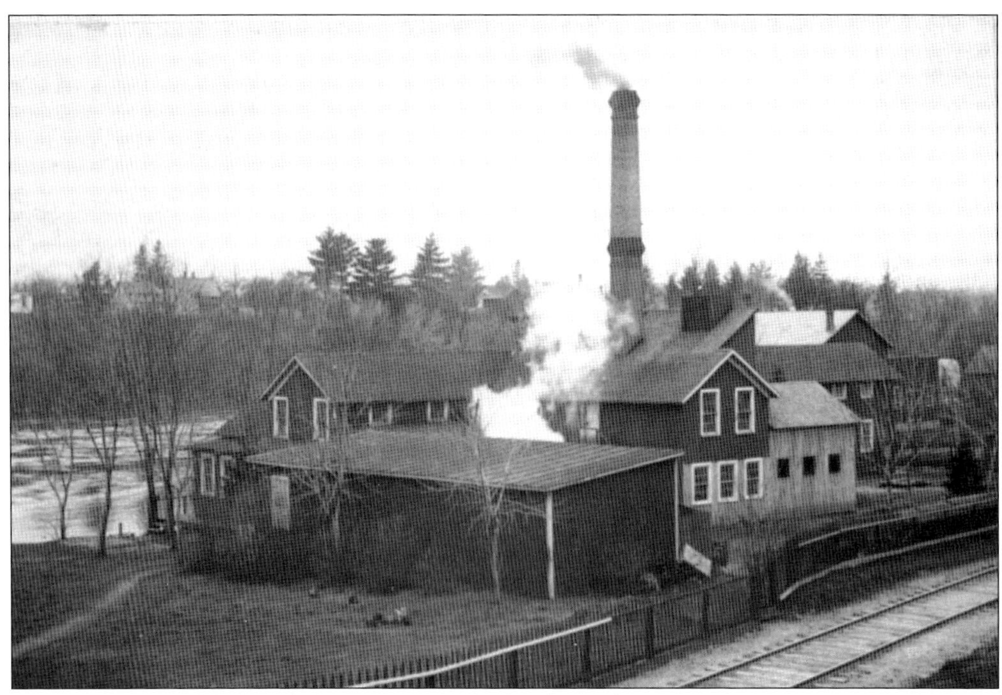

In 1883, W.A. Reddick built his factory on the banks of the St. Joseph River, just north of the Main Street Bridge near the Big Four Railroad line. A fire destroyed the factory in 1914. A year later, William Reddick's nephew, Benjamin Smith, bought his uncle's interests in the industry and opened a new factory on Second Street under a new name, Michigan Wire Goods.

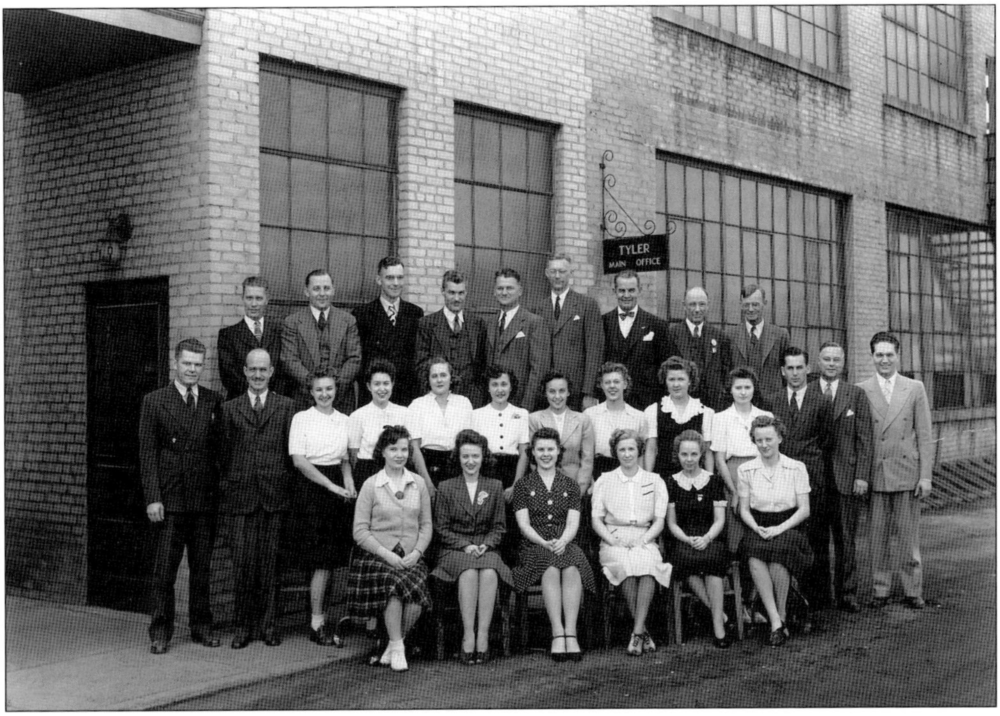

In 1927, Jerry Tyler founded the Tyler Sales Fixture Corporation, later Tyler Refrigeration Company, in Muskegon Heights. The business produced refrigerator display racks for retail stores. In 1932, Tyler moved to Niles and purchased the Dry-Kold Refrigeration Company to expand his product line. The photograph shows the 1943 Tyler office staff. Jerry Tyler is visible in the third row, seventh from the left.

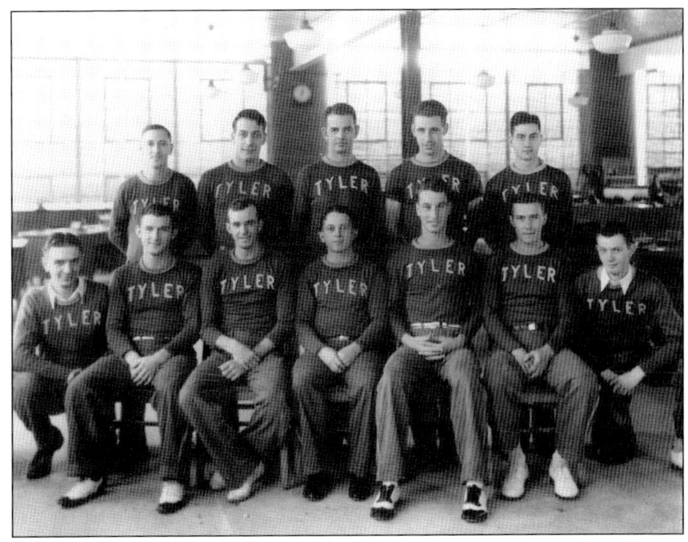

This photograph shows the 1935 Tyler Refrigeration softball team. Jerry Tyler is standing in the second row, third from left. He remained president of the company until his tragic death in the 1946 Chicago LaSalle Hotel fire. His brother, Robert Tyler, led the company until 1969. Tyler Refrigeration saw many monumental achievements in the commercial refrigeration industry, including the first walk-in freezers and open, self-service frozen food displays.

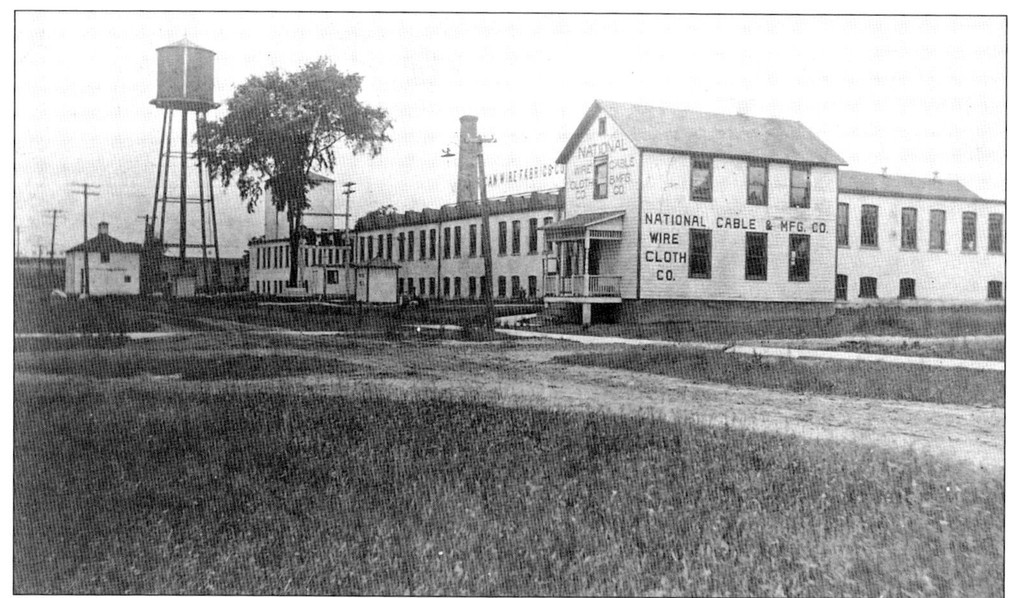

In the early 1900s, William Harrah moved to Niles to manage the National Wire Cloth Company. In 1906, he took charge of a department that produced lightning rods from round cables of "pure standard wire." A year later, that department incorporated as a new business under the name National Cable and Manufacturing Company, with Harrah as president. In 1913, after acquiring Cook Standard Tool Company, the business was renamed National Standard.

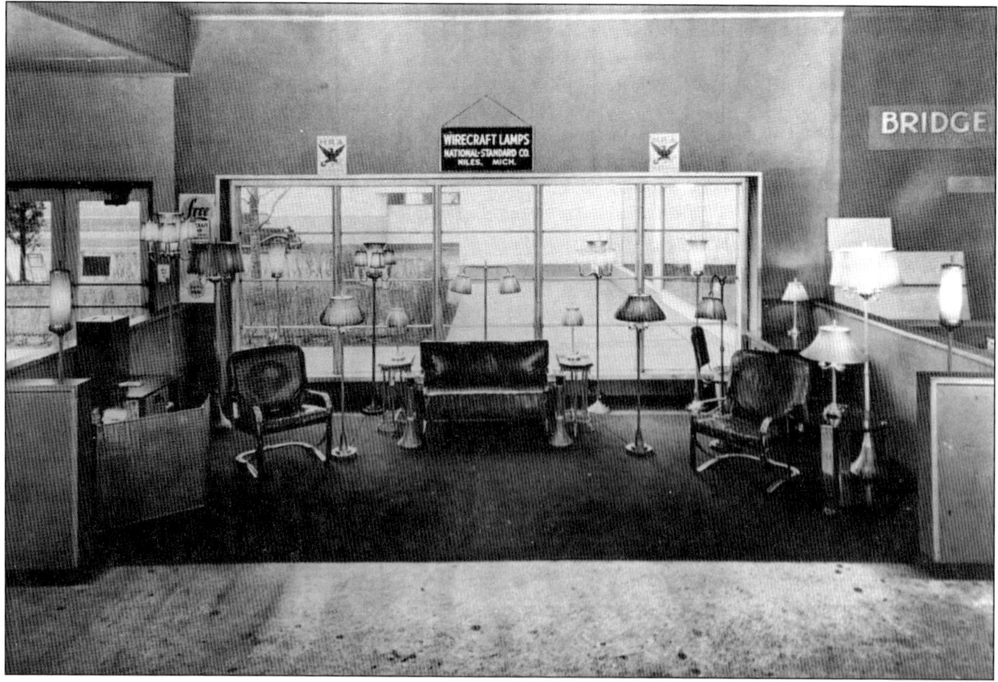

National Standard produced a variety of wire products, mostly for automobile tires. The company did develop other items, including baskets and lamps made of woven wire. Produced from 1930 to 1935, the home décor line was featured at the 1933–1934 Century of Progress Exposition in Chicago. The photograph shows the display room at the fair.

Kawneer changed the appearance of Main Street America. While living in Kansas City, founder Francis Plym developed an innovative system for framing glass windows. Made of rolled copper, the frames lasted longer and provided more support than wood frames. Plym received a patent for the design and started the Kawneer Company in 1906.

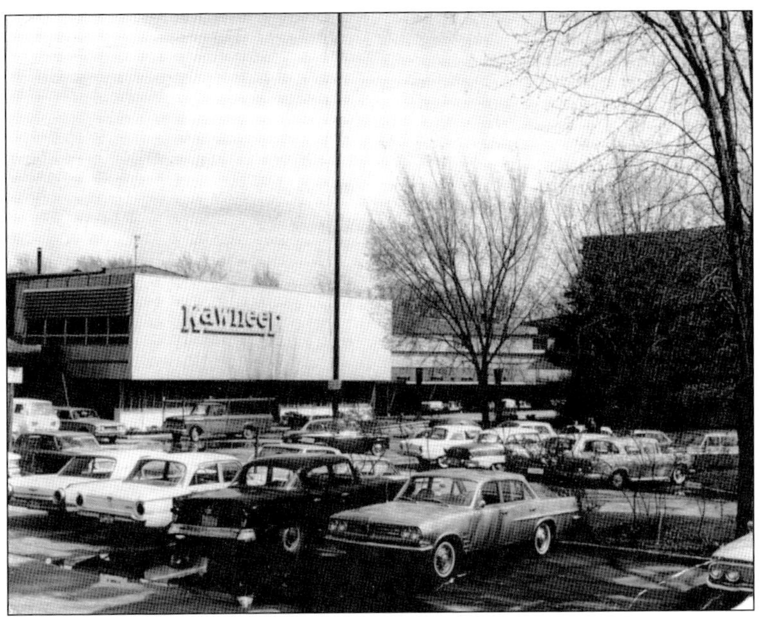

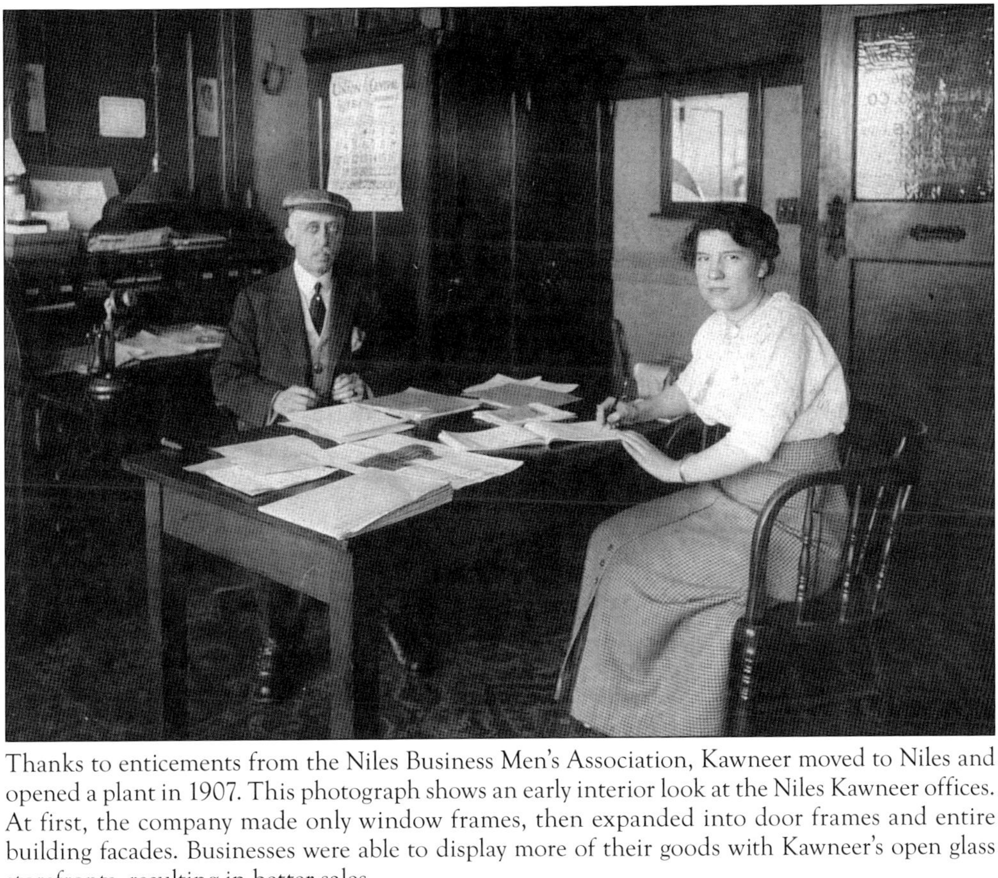

Thanks to enticements from the Niles Business Men's Association, Kawneer moved to Niles and opened a plant in 1907. This photograph shows an early interior look at the Niles Kawneer offices. At first, the company made only window frames, then expanded into door frames and entire building facades. Businesses were able to display more of their goods with Kawneer's open glass storefronts, resulting in better sales.

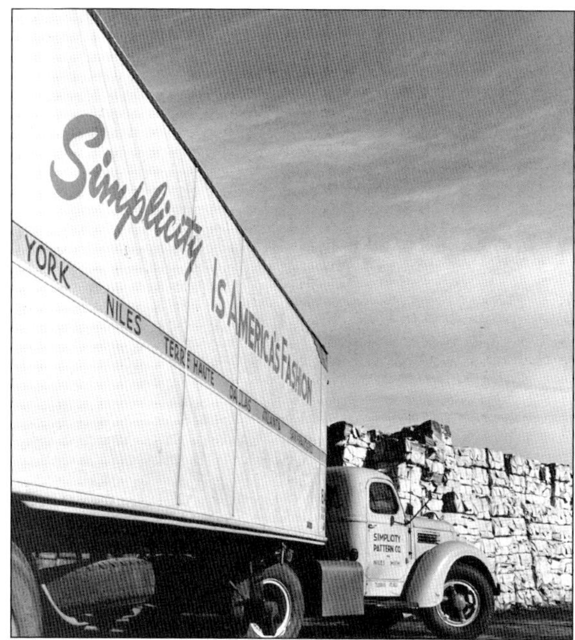

Located in Niles from 1931 to 2007, Simplicity Pattern Company was one of the area's largest employers. Founded in 1927, Simplicity started as a small business in New York with the philosophy of offering easy-to-follow patterns at an affordable price—just 15¢. The company quickly became a household name. In December 1931, Simplicity expanded, opening a plant in Niles, which served as the company's principal manufacturing center.

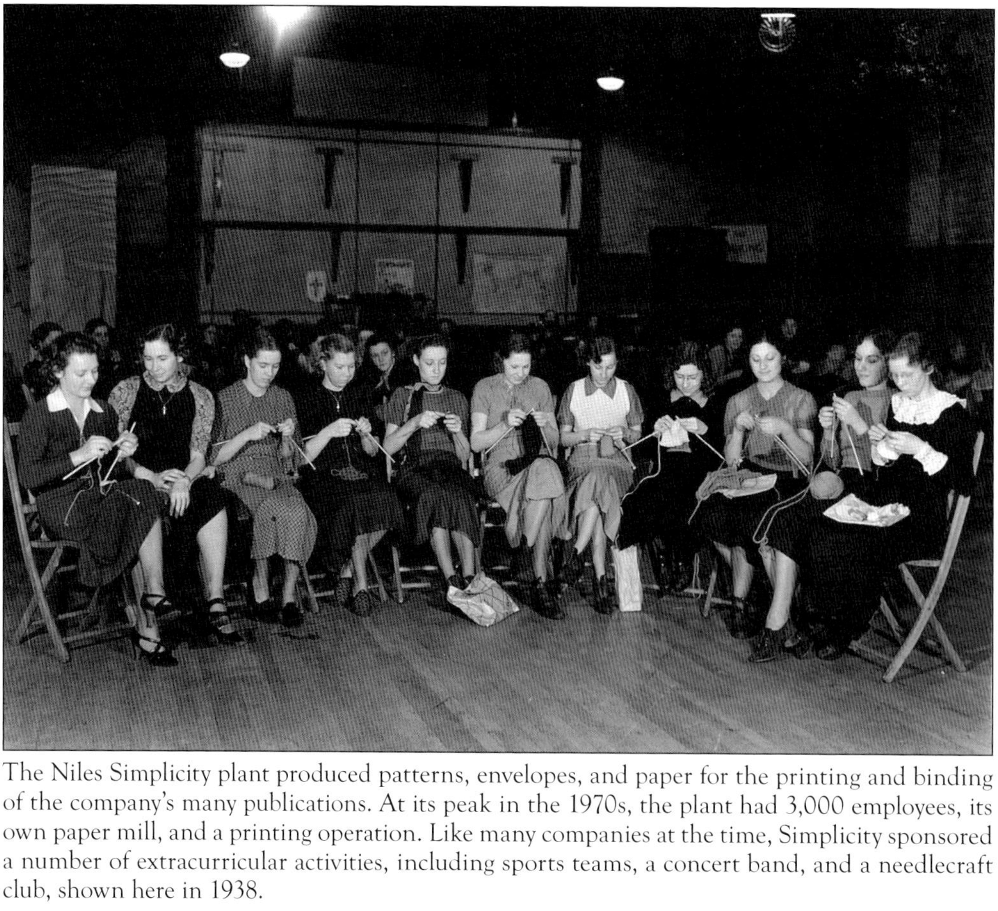

The Niles Simplicity plant produced patterns, envelopes, and paper for the printing and binding of the company's many publications. At its peak in the 1970s, the plant had 3,000 employees, its own paper mill, and a printing operation. Like many companies at the time, Simplicity sponsored a number of extracurricular activities, including sports teams, a concert band, and a needlecraft club, shown here in 1938.

Five

Honor and Service

"Thank God for Michigan!" Pres. Abraham Lincoln is said to have exclaimed when Michigan regiments quickly filled their quotas at the outbreak of the Civil War. In Niles, men rushed to enlist at Kellogg Hall, located downtown at Third and Main Streets. Nearly 1,000 soldiers were trained at Camp Barker, established at the fairgrounds east of town. Local African American men served with the US Colored Troops. Women contributed as well, some serving as nurses while others organized aid societies to sew flags, gather supplies, and raise money for troops.

Niles residents likewise answered the call during the Spanish-American War and in the major global conflicts of the 20th century. During World War I, LaRue Messinger became the first casualty from Niles when he was killed on a battlefield in France. Women served overseas with the Red Cross and some in combat roles. In World War II, Nilesites fought in all branches of the military and on both major fronts. With men away at war, women entered the workforce in unprecedented numbers. At home, residents planted victory gardens, purchased war bonds, donated money and supplies, and organized civil defense units. The Korean and Vietnam Wars once again called up Niles troops overseas.

Returning veterans helped shape civic life. Civil War veterans formed a local chapter of the Grand Army of the Republic (GAR), which reenacted battles, marched in parades, and fostered fellowship among those who had served. Memorials were built at Riverfront Park and Donovan Smith Park. The LaRue Messinger American Legion Post continues to organize the annual Memorial Day parade. The local Silverbrook Cemetery decorates the graves of soldiers who served as far back as the Revolutionary War.

Though the 17th-century Fort St. Joseph saw relatively little combat, both French marines and British soldiers were once stationed there. Archaeological finds—including buttons, weapons, and musket balls—confirm their presence. The British re-garrisoned the fort briefly during the American Revolution, making it one of the few sites in Michigan directly tied to the nation's struggle for independence.

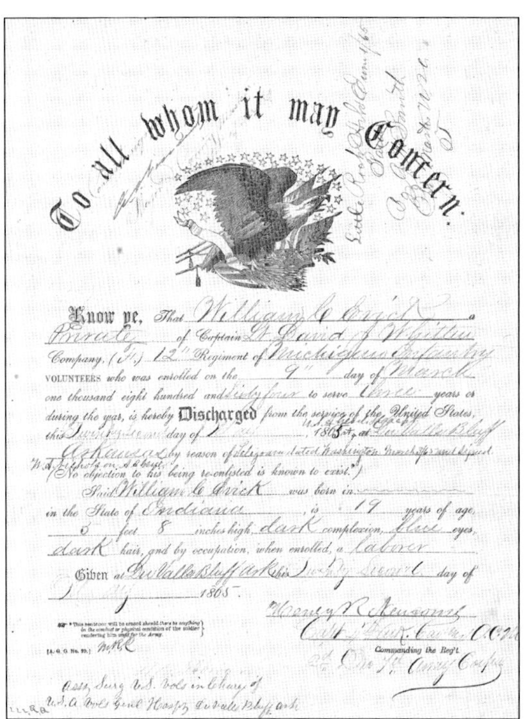

To all whom it may Concern.

Know ye, That William C. Evick a Private of Captain Lt. David J. Whitten Company, (F) 12th Regiment of Michigan Infantry VOLUNTEERS who was enrolled on the 9th day of March one thousand eight hundred and sixty four to serve three years or during the war, is hereby Discharged from the service of the United States, this Twenty Second day of [illegible], 1865, at DuValls Bluff Arkansas by reason of [illegible]

(No objection to his being re-enlisted is known to exist.*)

Said William C. Evick was born in [illegible] in the State of Indiana, is 19 years of age, 5 feet 8 inches high, dark complexion, blue eyes, dark hair, and by occupation, when enrolled, a laborer

Given at DuValls Bluff Ark this Twenty Second day of May 1865.

Henry R. Newsome

Commanding the Reg't.

*This sentence will be erased should there be anything in the conduct or physical condition of the soldier rendering him unfit for the Army.

[A. G. O. No. 99.]

In 1861, Francis Quinn received an authorization from Governor Blair to raise the 12th Michigan Infantry and rendezvous in Niles. 1,000 soldiers trained at Camp Barker, constructed at the old fairgrounds just east of downtown. On March 19, 1862, the unit left for the Battle of Shiloh in Tennessee. Colonel Quinn's disastrous actions during the battle led to many casualties and resulted in his resignation. Over 500 men from Niles and Cass Counties served in the 12th Michigan Infantry, including William Evick of Company F, whose discharge papers are seen here.

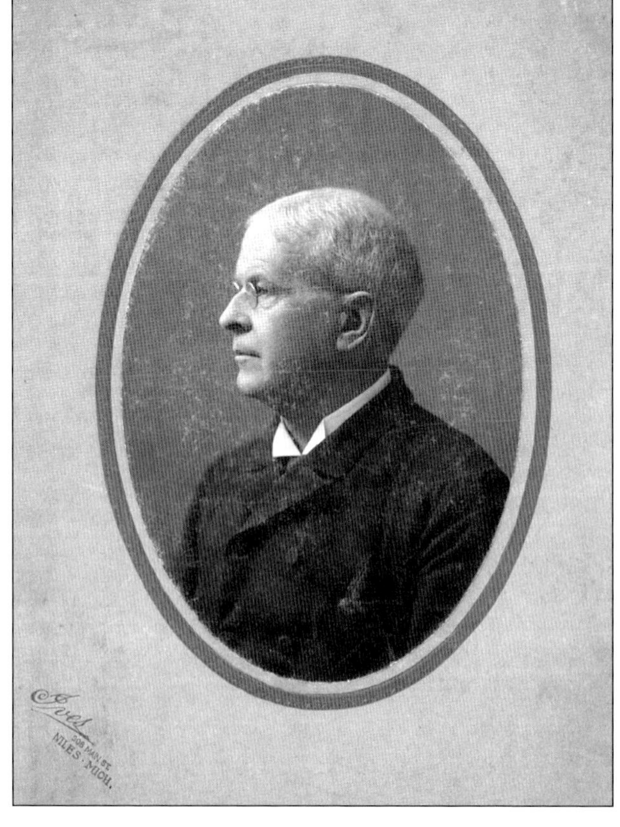

Edward Bacon of Niles served in the 6th Michigan Infantry during the Civil War. He earned the rank of colonel. Berrien County's earliest Civil War company formed in Niles and became part of the 2nd Michigan Infantry Regiment. Around 1,000 men enlisted from Niles and the surrounding area, serving in various units, including the 2nd, 6th, 12th, and 19th Michigan Infantries as well as the 1st Michigan Colored Infantry.

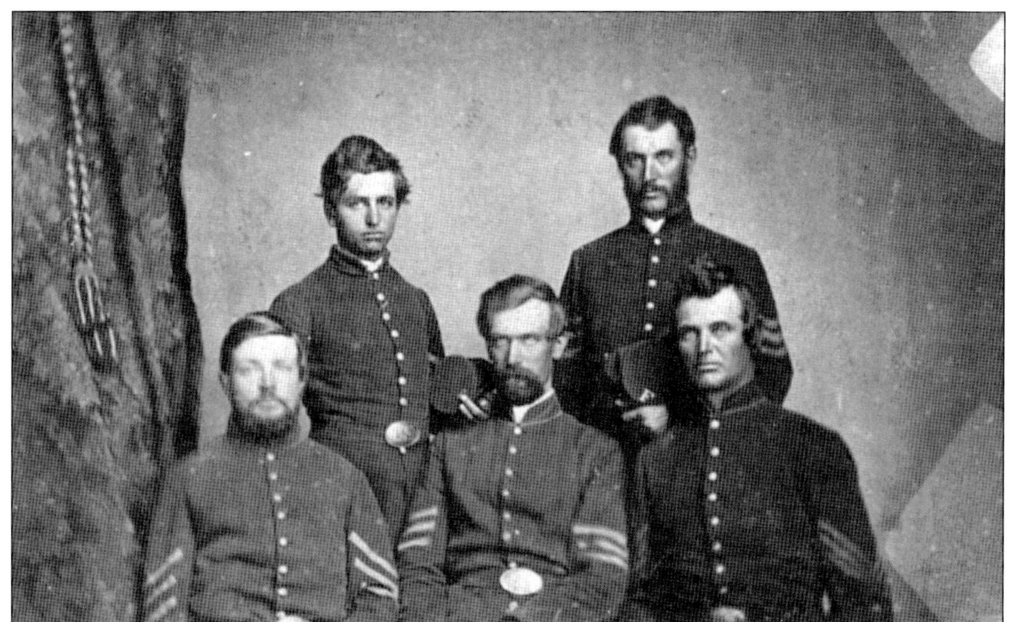

The 25th Michigan Infantry was organized in Kalamazoo and mustered in on September 22, 1862. Three of the ten companies were from the area, including Company F of Niles. This photograph shows sergeants in the company. From left to right are (seated) Henry Bond, Irving Paddock, and Don Clark; (standing) Charles Woodruff and Henry B. Adams. All these men were from Niles, except Paddock, who was from Three Oaks.

Noah Cain served in Company G of the First Michigan Sharpshooters during the Civil War. Cain enlisted on June 20, 1863, and died of disease in Virginia on August 17, 1864. Around 30 men from Niles served in the First Michigan Sharpshooters, several as officers. To join, recruits had to pass a test of marksmanship. Company K of the regiment was entirely made up of Anishinaabe enlistees.

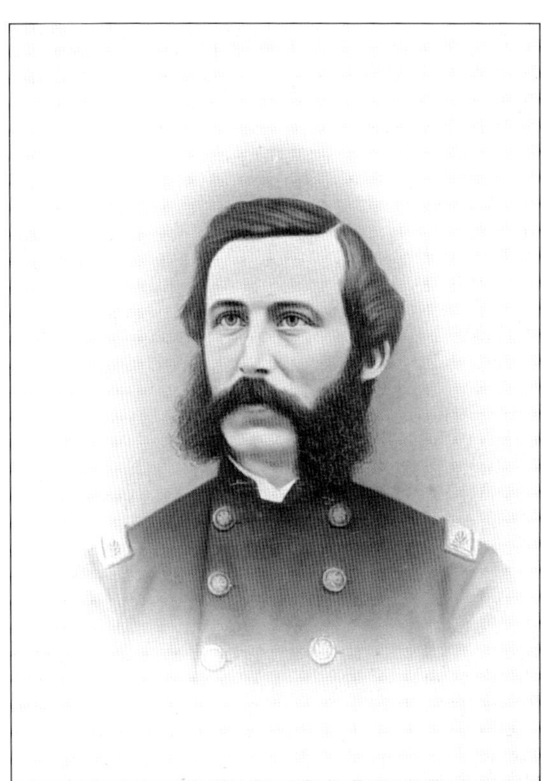

At the start of the Civil War, Niles native Eli Griffin organized the Berrien County Rifles, designated Company A of the 6th Michigan Infantry. Griffin served as captain until he was injured in July 1863. He returned to active duty in October as a major in the 19th Michigan Infantry, organized at Dowagiac. Mortally wounded in battle on June 15, 1864, Griffin died the next day, leaving a wife and young son.

Evan Bonine, seated front row center, is shown with a group outside a tent with a sign noting "Surgeon's Headquarters 3rd Division Hospital, 9th Army Corps." Born in Richmond, Indiana, in 1821, Dr. Evan Bonine moved to Niles with his wife, Evaline, in 1858 and started a medical practice. At the outbreak of the Civil War, Bonine was appointed surgeon of the 2nd Michigan Infantry. He advanced to chief of staff and division surgeon.

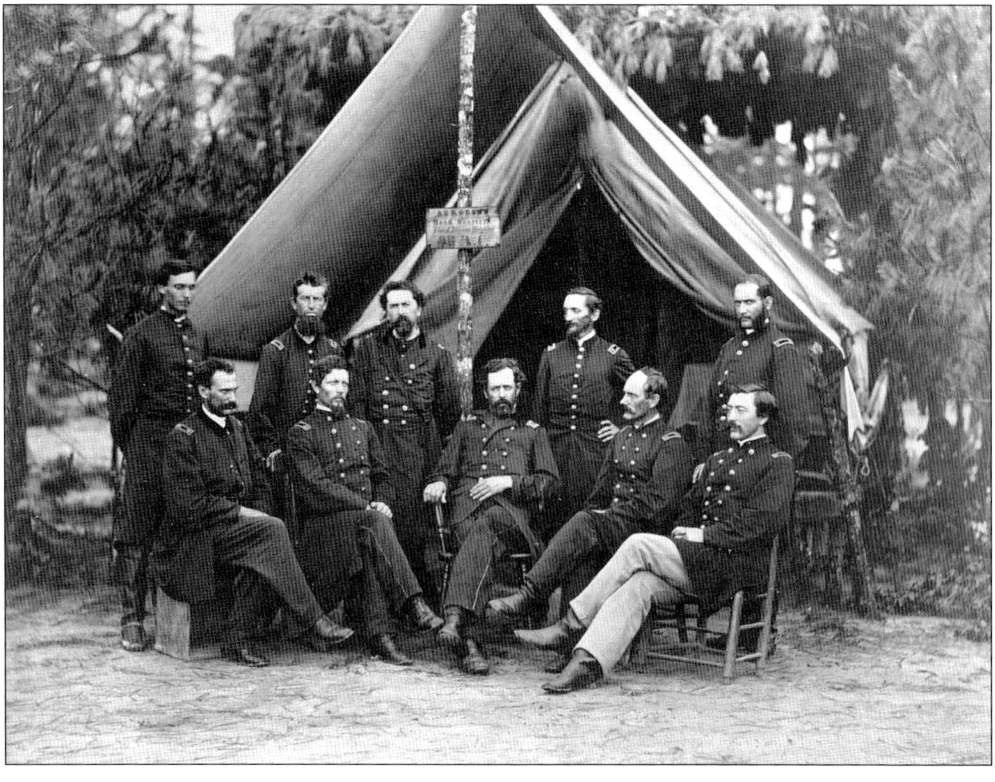

On the home front, Evaline Bonine, pictured here, participated in the local Ladies' Aid Society. After the war, Evan Bonine practiced as a physician in Niles and participated in local and state government. Their son Frederick was born in 1863 and followed in his father's footsteps, becoming an eye doctor and serving in local politics.

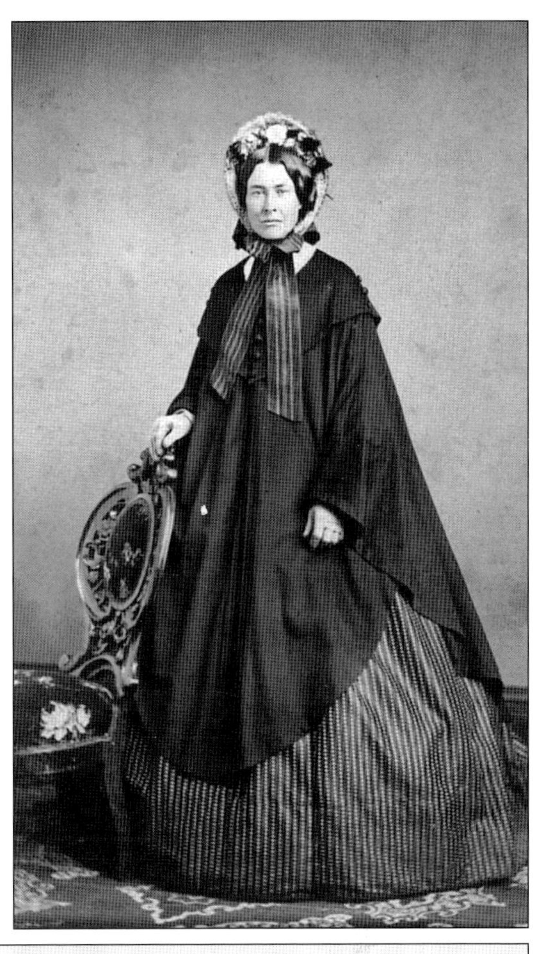

In August 1863, just months after President Lincoln issued the Emancipation Proclamation—which allowed Black men to enlist in the military, albeit in segregated units—Governor Blair received permission to organize the 1st Michigan Colored Infantry Regiment, which became the 102nd US Colored Troops. Ninety-nine men from Berrien and Cass counties joined the regiment. Many from Niles served in Company G, including William Powers, whose papers are shown here.

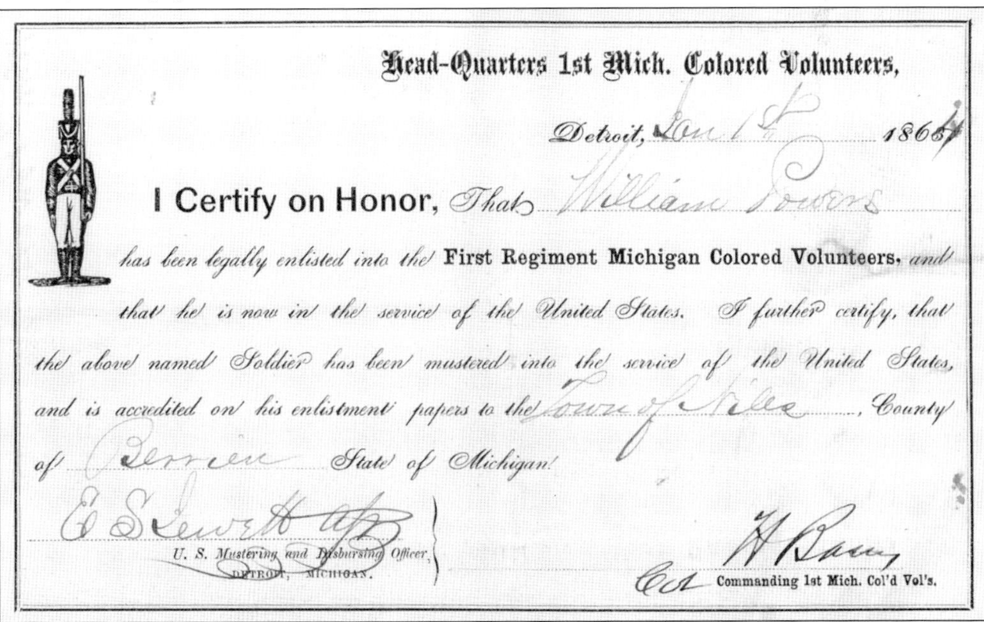

Head-Quarters 1st Mich. Colored Volunteers,

Detroit, Jan 1st 1864

I Certify on Honor, That William Powers

has been legally enlisted into the First Regiment Michigan Colored Volunteers, and that he is now in the service of the United States. I further certify, that the above named Soldier has been mustered into the service of the United States, and is accredited on his enlistment papers to the Town of Niles, County of Berrien State of Michigan.

E S Leavitt Capt
U. S. Mustering and Disbursing Officer, Detroit, Michigan.

H Barns
Col Commanding 1st Mich. Col'd Vol's.

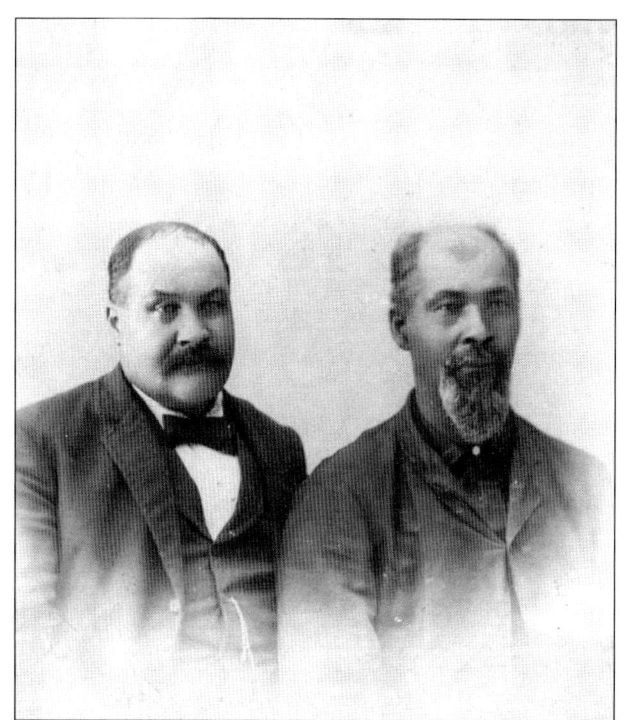

Edward Finley (left), son of Pasquel and Sarah Finley, served in Company I of the 102nd US Colored Troops. He enlisted as a private in Niles on January 21, 1864, and was discharged as a sergeant in Charleston, South Carolina, on September 30, 1865. His brother Greenville (right) did not serve in the Civil War, although another brother, Richard, also served in the 102nd, Company E.

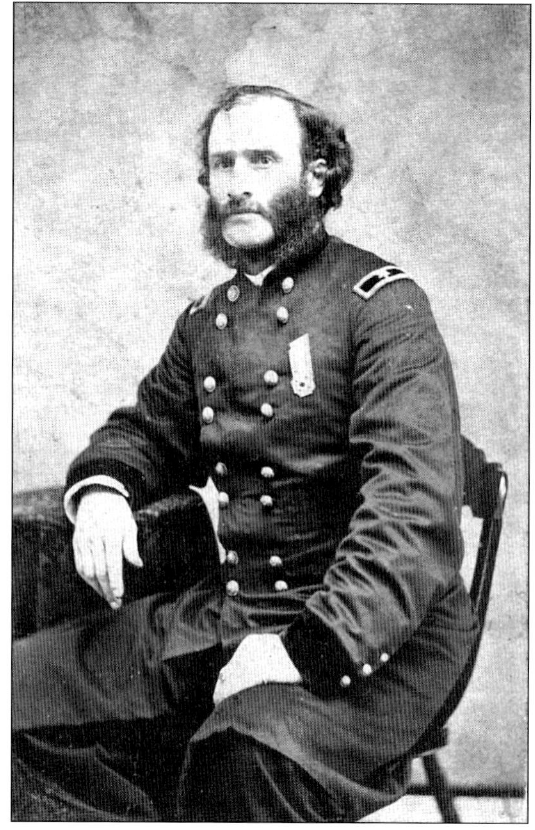

In 1862, Governor Blair authorized Henry Morrow (pictured here) to raise the 24th Michigan Infantry. The regiment mustered into service on August 15, 1862, and selected Morrow as colonel. In October 1862, the 24th Michigan joined the Iron Brigade. Named for their tough reputation, the Iron Brigade consisted of units from Wisconsin, Indiana, and Michigan. They fought in several major battles, including Antietam, Fredericksburg, Chancellorsville, and Gettysburg. By mid-1864, the 24th included 70 recruits from Berrien County. Morrow was promoted to brigadier general.

In 1860, Morrow married Isabella Graves of Niles. Isabella's brother, Frank Graves, also served in the war as a colonel in the 8th Michigan Infantry. Graves died during the Battle of the Wilderness in 1864. The Niles GAR (Grand Army of the Republic) post, organized in 1882, is named after him.

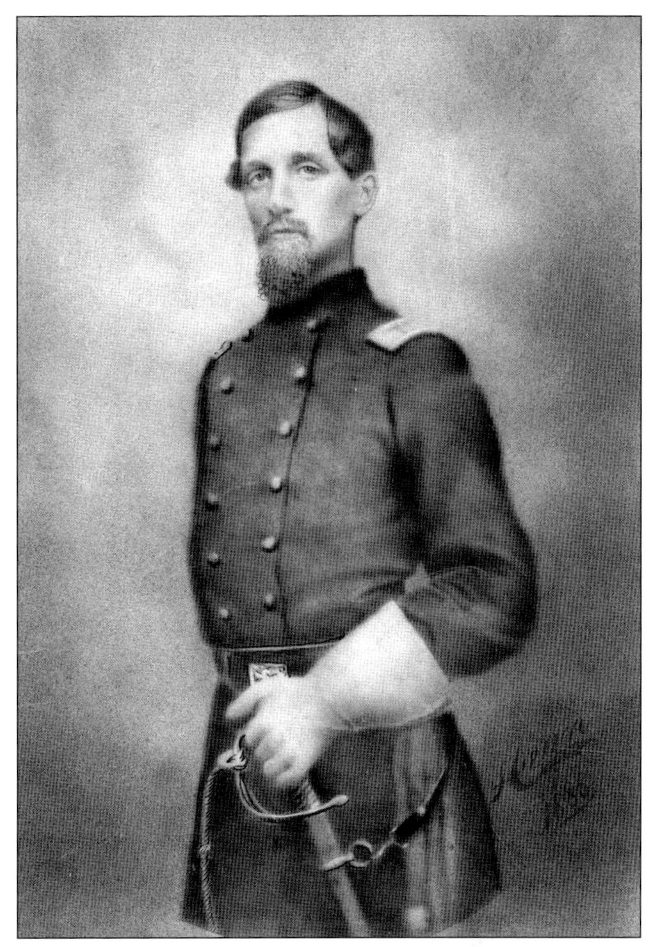

The Grand Army of the Republic formed as a national organization for Union veterans. In Niles, a short-lived GAR post, named after Eli Griffin, was formed in 1866 but disbanded after a few years. The Frank Graves Post, No. 64, was organized in 1882. The group held meetings and reunions through the early 1920s. This photograph shows members of the Frank Graves post at their hall on Main Street.

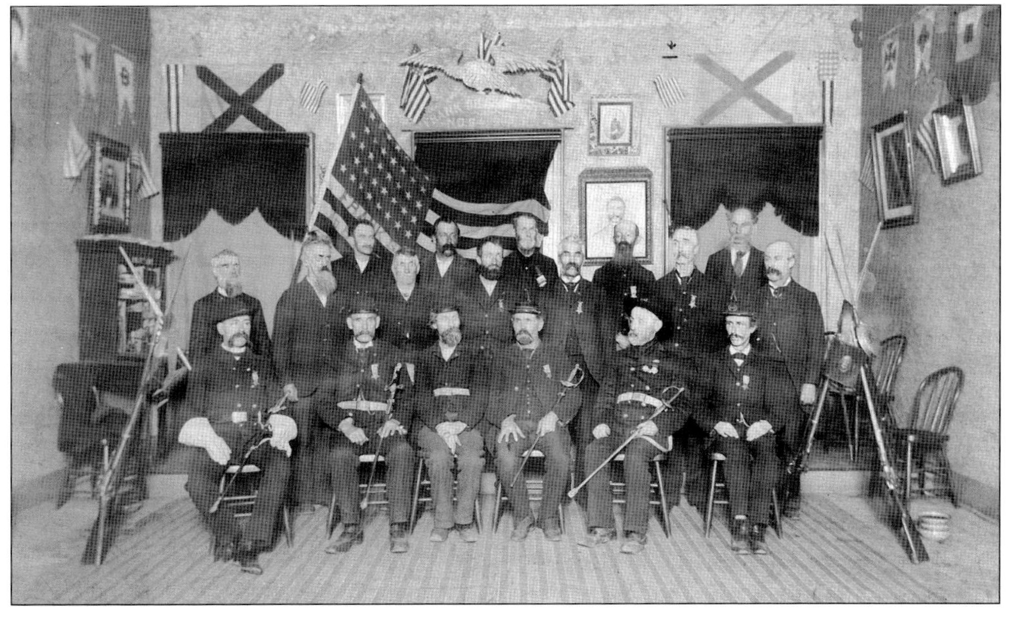

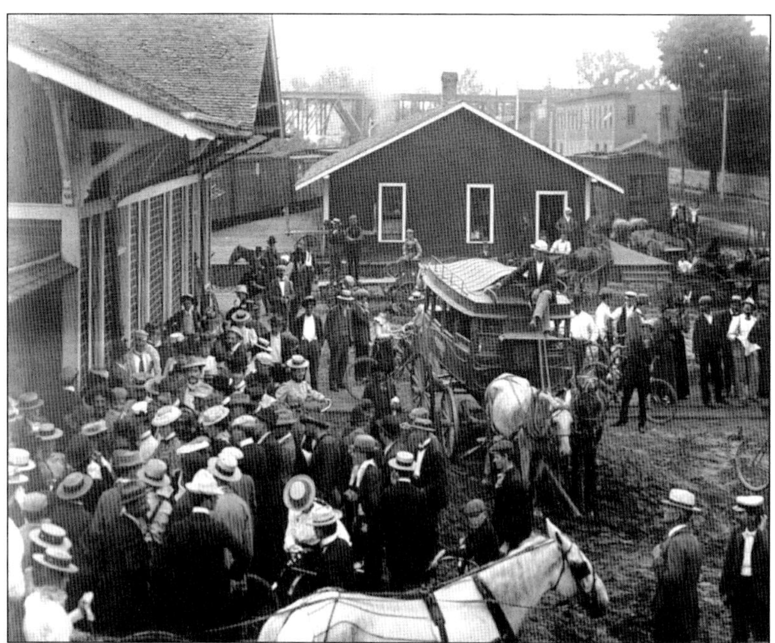

A crowd at the platform of the Big Four Railroad depot greets three Niles boys returning from the Spanish-American War: Horace B. Corell, George Edward Corell, and Lewis Johnson. The *Niles Weekly Mirror* notes, "At the sight of them a mighty cheer arose. Before the train had come to a full stop the boys were on the platform with their military baggage."

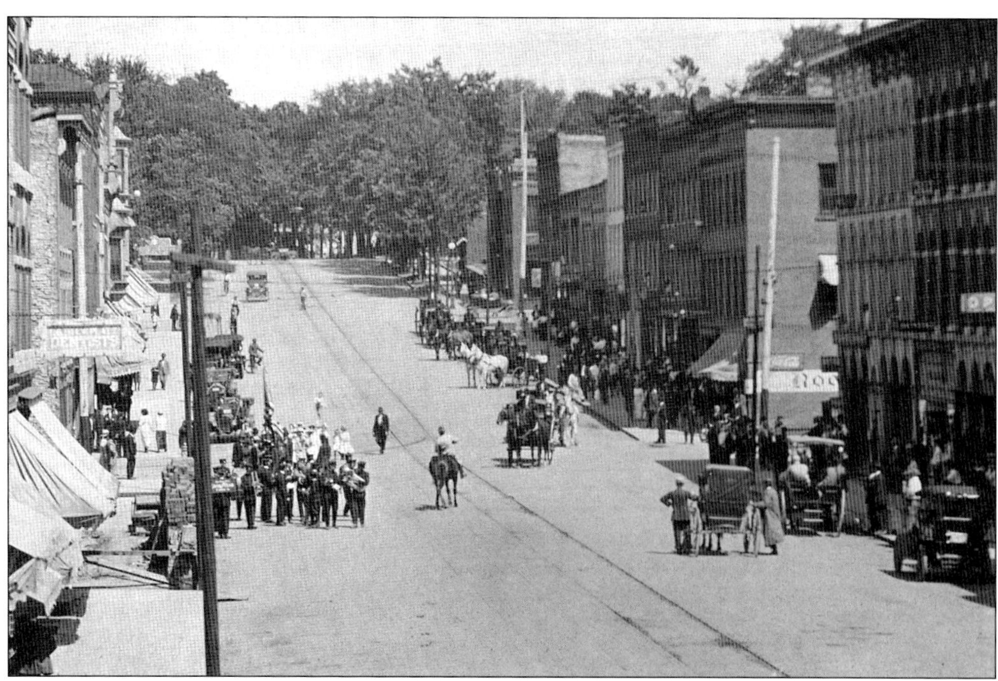

A Decoration Day parade in 1914 makes its way east on Main Street. Spanish-American War veteran George E. Corell is on horseback near the interurban rail tracks. Decoration Day began as an occasion to honor the graves of Civil War veterans. After World War I, Memorial Day became the more common name, honoring fallen service members from all wars.

LaRue Messinger fought in Company K of the 6th Infantry Division during World War I. During the Battle of Saint-Mihiel, France, on September 12, 1918, La Rue was killed in action while manning a trench mortar. He was the first person from the Niles area to die in combat during the war. American Legion Post 26 was named after the young soldier.

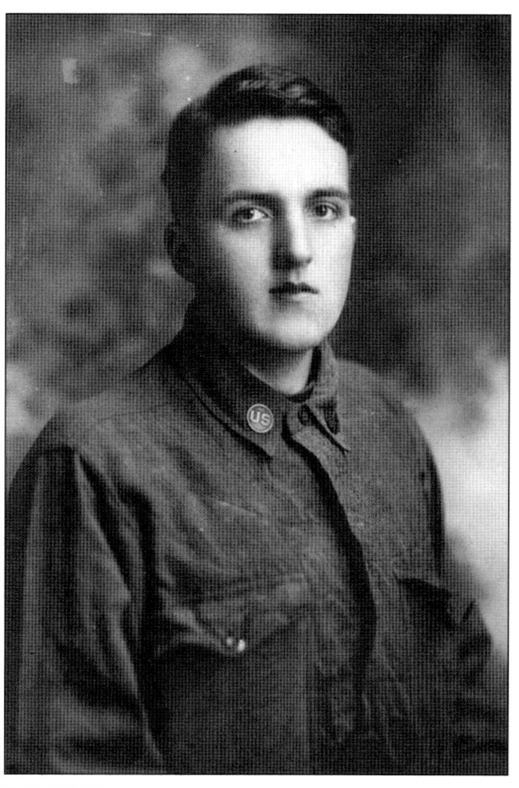

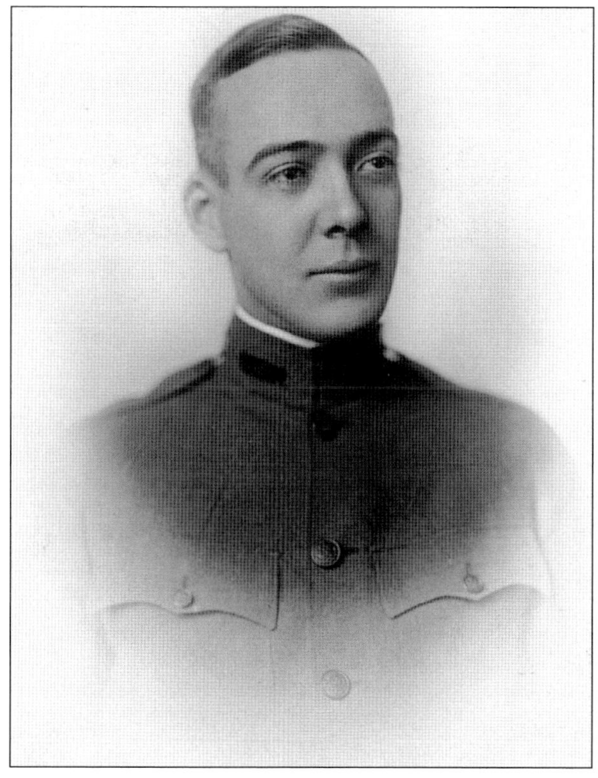

During World War I, Frederick Kompass enlisted in May 1917 and served in France, Company B, 38th US Infantry, earning the rank of lieutenant. He was a charter member and commander of LaRue Messinger Post of the American Legion. After the war, he worked in the furniture manufacturing business owned by his father, Rudolph, and Matthew Stoll, until it was sold in 1928. Frederick then founded Kompass Products, producing novelty furniture and post office furniture.

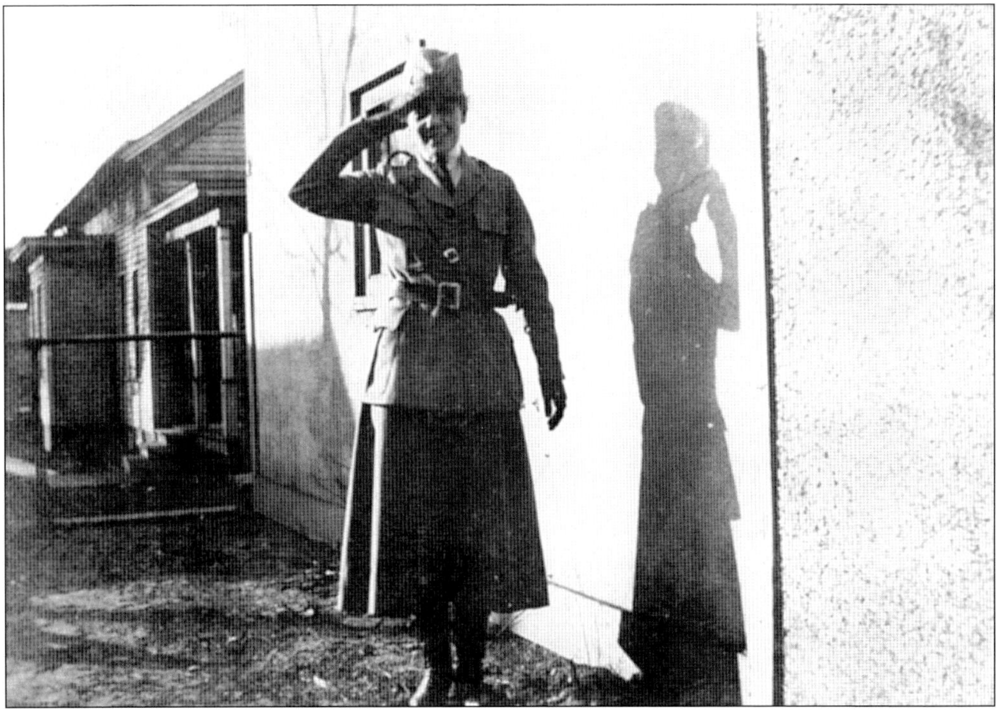

Pictured here is Lt. Gladys Forler, who, along with Allene French, organized a branch of the Red Cross Michigan War Motor Corps during World War I. Many women joined the corps, which supplied transportation to organizations and personnel connected with the war. While not a division of the military, the corps worked with the Michigan War Preparedness Board.

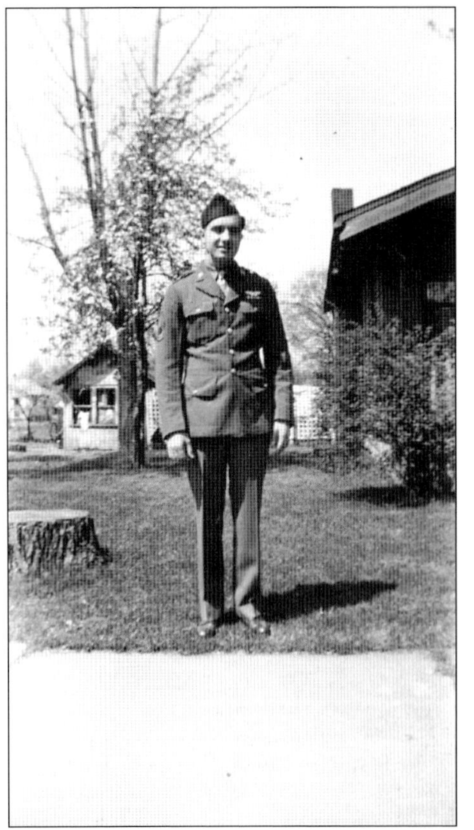

Long before he was a beloved Niles mayor, Frank Frucci enlisted in the US Army Air Force early in World War II. Stationed in the Pacific, he was in Pearl Harbor during the 1941 surprise attack and later helped found the Michigan Pearl Harbor Survivors' Association. He served as a gunner in a B-17 bomber during the Battle of Midway, earning the rank of master sergeant.

Harold Huttenstine served as a Navy seaman first class in the Pacific Theater during World War II. After the war, he returned to Niles and worked for National Standard. He also owned and operated Hutt's Rubber Stamp Shop for 26 years. As a civilian, he continued his service as a volunteer fireman for the Niles Township Fire Department and as chairman of the Niles Chapter of the March of Dimes.

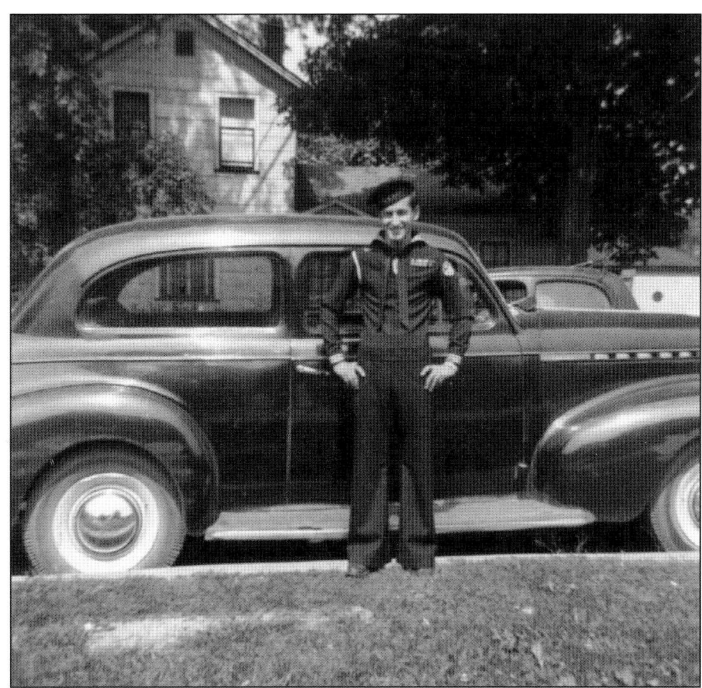

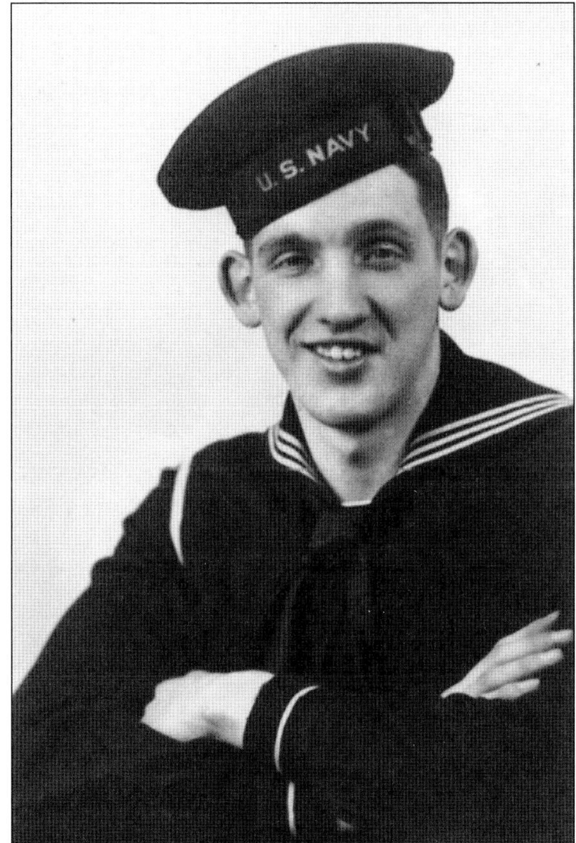

Louis Drolet enlisted in the US Navy V-5 shortly after graduating from Niles High School in 1943 and reported to the Navy College Training program at the University of Notre Dame. V-5s were high school graduates who took college classes for two years while training as naval aviators. In July 1943, the V-5 combined with the V-7 program to form the Navy V-12.

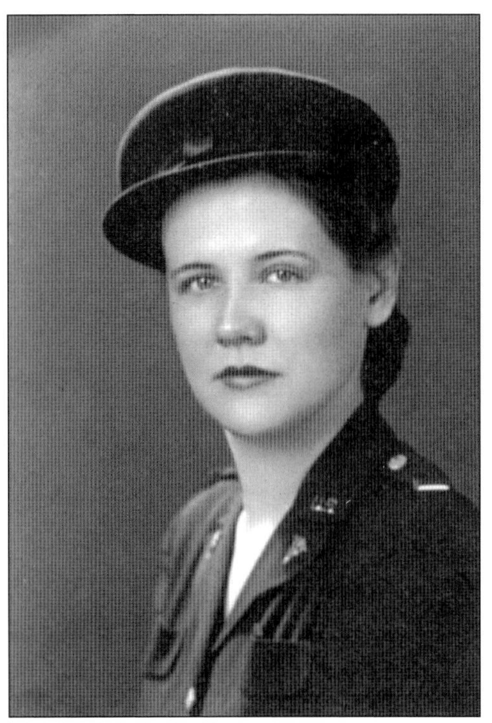

Irma Holmdahl Hershey first enlisted in the US Army in 1940 as a nurse anesthetist. She later re-enlisted with the 27th Evacuation Hospital of the Seventh Army. She served in Europe and achieved the rank of first lieutenant, earning many honors. She married Dr. Noel Hershey and continued work as a nurse anesthetist at Pawating Hospital. Lieutenant Hershey is buried at Arlington National Cemetery.

During World War II, local industries shifted production to help the war effort. Women entered the workforce in larger numbers than ever before. Area factories, including Kawneer, Simplicity Pattern, and Tyler Refrigeration, employed women for various jobs. This 1943 photograph shows several women working at Tyler Refrigeration Company's Plant No. 3. During the war, Tyler produced products for the military, including refrigerators and cargo trailers.

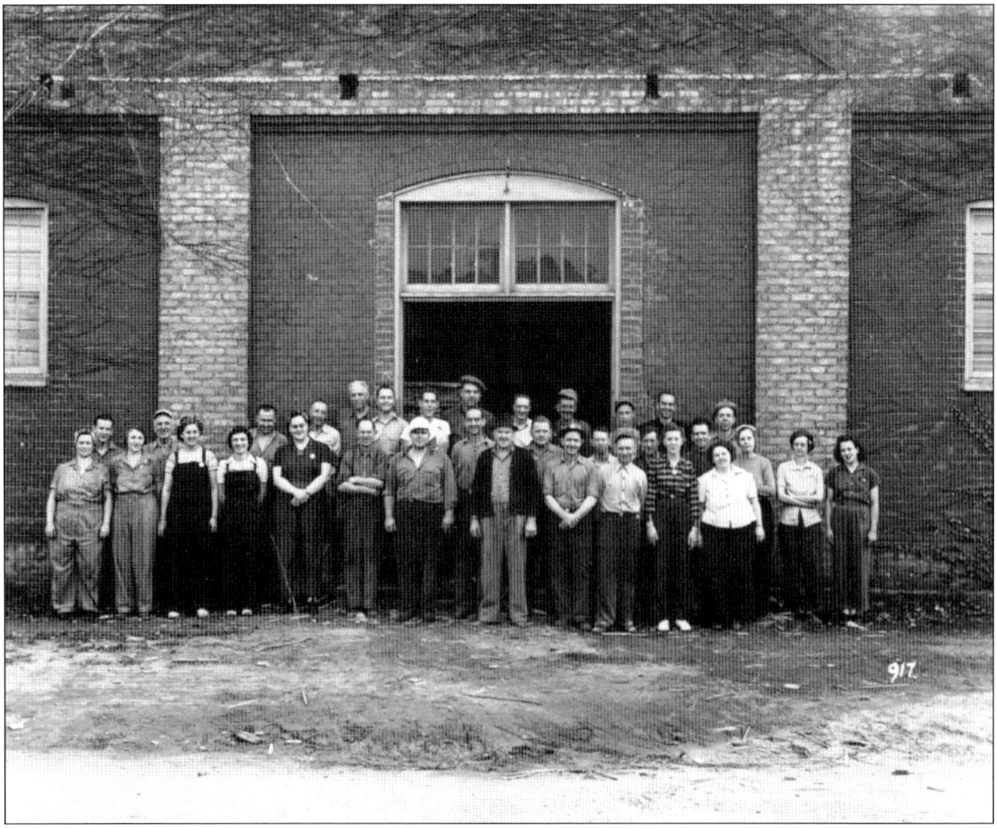

Lillian Luckey of Niles is pictured working on an airplane at the Marine Corps Air Station in El Toro, California. Luckey joined the Marines in 1943. During her service, she worked as an aircraft mechanic and also played softball for the base's women's team. Luckey returned to Niles after her service. In 1946, the All-American Girls Professional Baseball League (AAGPBL) recruited her to pitch for the South Bend Blue Sox team.

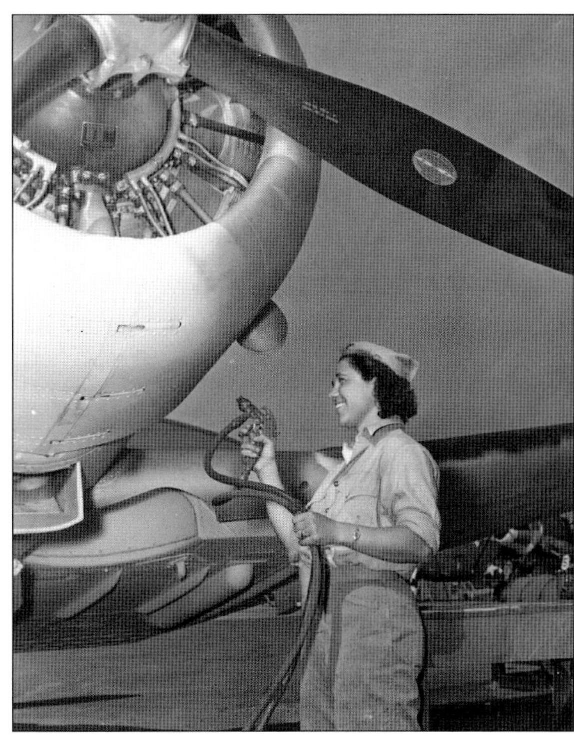

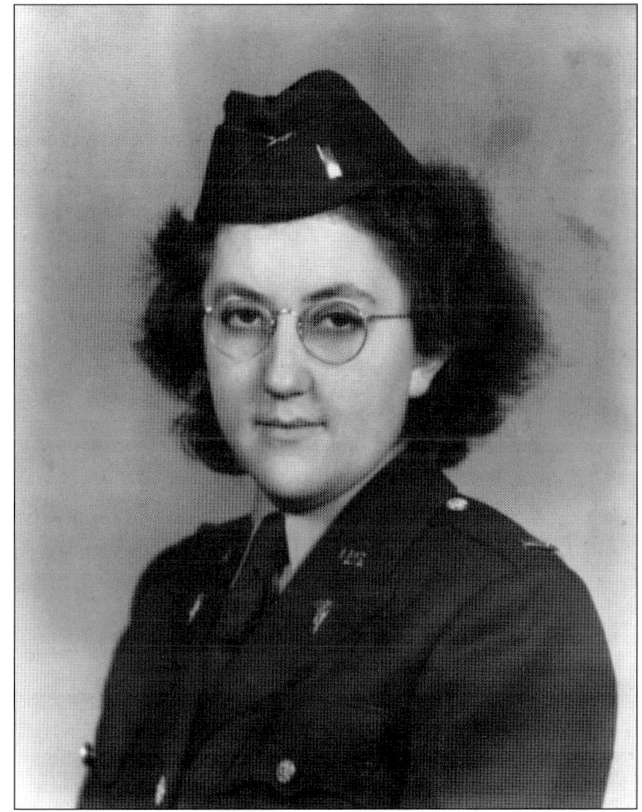

Dorothy L. Bolinger served as a first lieutenant nurse in the US Army during the Korean War. Her father, Frederick, served in World War I, and her brother Frank served in World War II. Bolinger cared for as many as 1,000 battle casualties in a semimobile hospital equipped to handle only 400. The hospital units could be located within a mile of the front lines, well within earshot of the fighting.

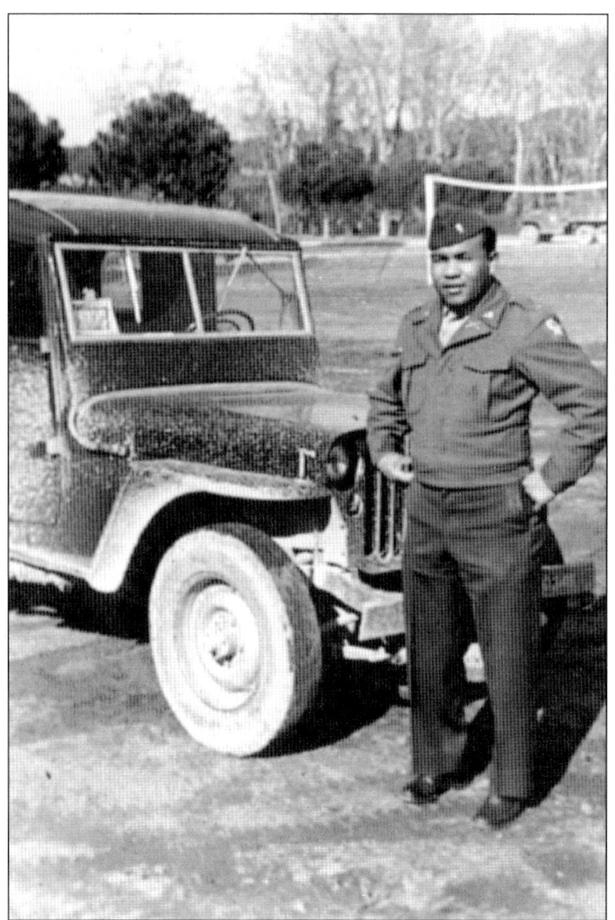

Seth D. Finley was a career military man who spent 20 years in the Army. He served in Italy and Germany during World War II, where he was promoted to first lieutenant. In Korea, he earned the rank of captain with the 24th Infantry Regiment. He retired as a lieutenant colonel on Nov. 30, 1962.

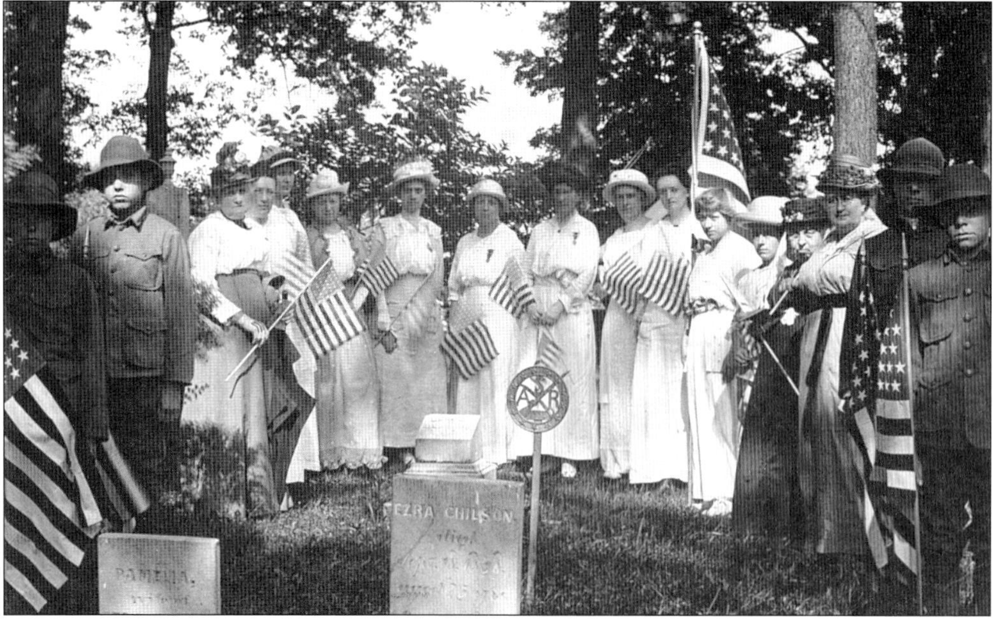

In 1915, members of the Fort St. Joseph Chapter, Daughters of the American Revolution, with assistance from a local Boy Scout troop, placed a bronze marker at the grave of Revolutionary War veteran Ezra Chilson in Silverbrook Cemetery. Born in 1762, Chilson served in the Revolutionary War out of Massachusetts. He moved to Niles around 1830. His son Hiram, also buried at Silverbrook Cemetery, served in the War of 1812.

Six

School Days, Sports, and Clubs

Soon after the Village of Niles was established, schoolteacher Titus B. Willard began holding primary classes in his log cabin. Secondary education was deemed essential to attract new settlers from the East Coast. The Niles Academy opened in 1836, instructing male pupils for a fee in math, geography, history, French, chemistry, and other sciences. The Niles Female Seminary opened later that same year. In 1838, townspeople were invited to observe examinations at the female seminary. All were suitably impressed with the students in this "backwoods country."

A free common school was opened at Sycamore and Third Streets during the 1830s, along with other neighborhood schools. Rural communities followed suit, establishing Pucker Street School, north of the city limits, Bertrand School, and others. Educational quality varied widely, depending on individual teacher ability. Schools suffered from overcrowding. To improve conditions, several districts united to build a larger school and employ qualified educators. Union School, a three-story Victorian structure, opened in 1856 at Broadway and Seventh Streets.

Before 1871, Niles schools were segregated. African American children attended the old common school, which was now in disrepair. In 1867, Calvin Wilson, Chair of the Colored Citizens of Niles, petitioned the city council for a better facility. Thanks to his efforts, a new school was built on Ferry Street. Isaac Burdine, an educator known across the region, was the first teacher. Wilson's daughter Charlotte ("Lottie") followed in her father's footsteps, advocating for education on the national stage.

When industry flourished and railroads expanded, Niles experienced a population boom. Eastside School welcomed students to a larger building. Ballard School opened, named for local historians Ralph and Mary Ballard. Another new building honored the nationally known author Ring Lardner. The Brandywine School District was organized to serve the surrounding townships. Other school districts, such as Howard in the Barron Lake area, consolidated with Niles Community Schools.

Schools offer more than reading, writing, and arithmetic. Sports, band, clubs, and a myriad of student organizations have enriched the lives of Niles students. To this day, the community continues to rally behind its schools and mascots, whether the Niles Vikings, Brandywine Bobcats, or a number of private institutions.

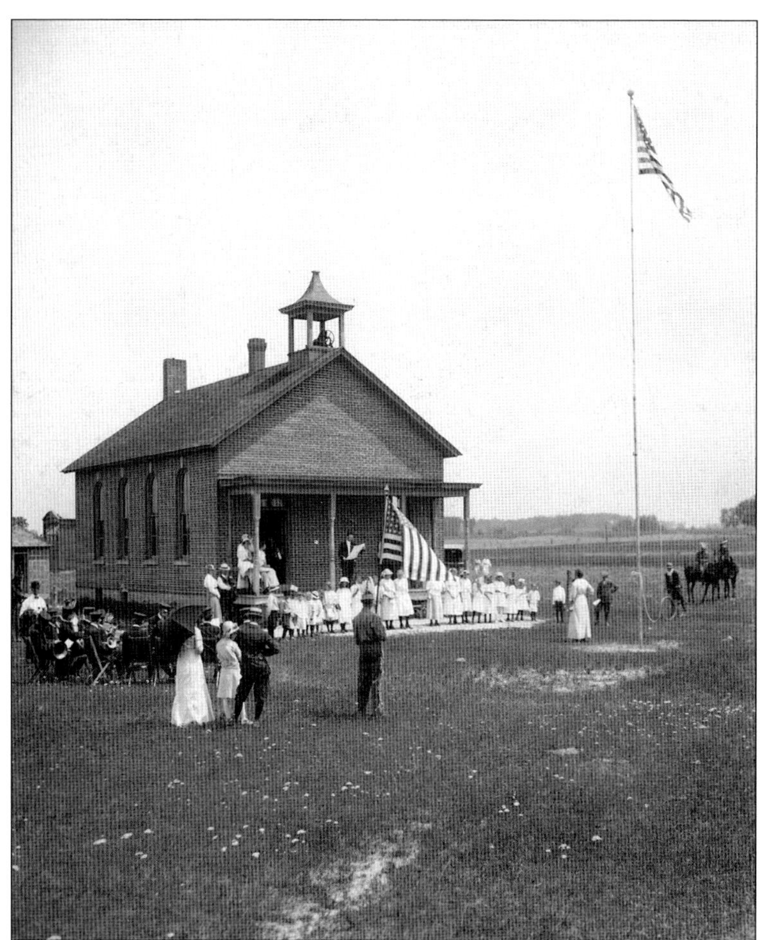

Bertrand built a new brick schoolhouse in 1899. This photograph likely shows the opening ceremonies with a flag raising, band, and students lined up to see their new school. The September 5, 1899, *Niles Daily Star* noted that "the new school is progressing fine and when it is finished, everyone is invited to an ice cream social."

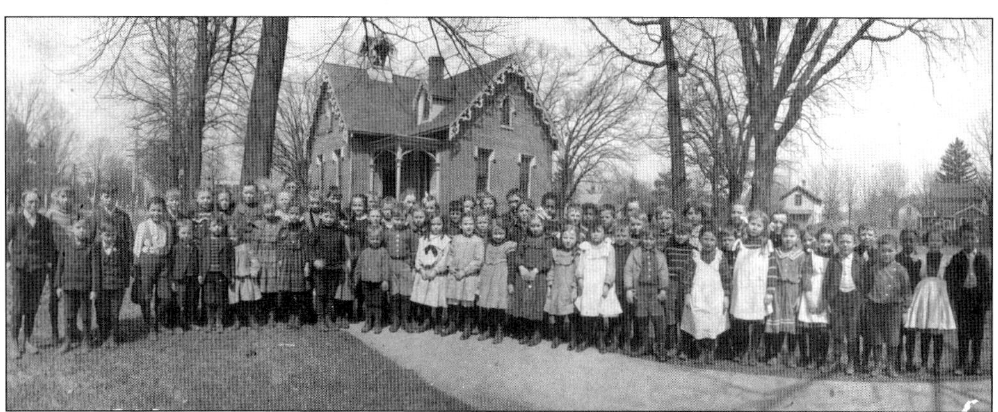

Smaller schools sprang up in each of Niles's four wards. The fourth ward is the area west of the St. Joseph River. The Fourth Ward School taught primary grades first through third, with one class for each grade. Here, students are pictured in front of the school at Lincoln Avenue and Broadway Street around the year 1909.

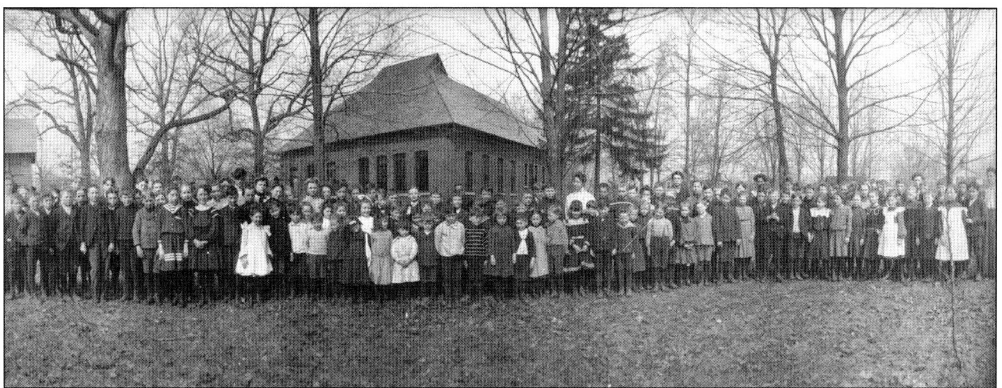

A new three-room school was built on Cherry and Fifth Streets in 1903. First called the Third Ward School, it eventually became known as Cherry Street School. Several grades were taught here, as seen in this photograph. Though there was talk of razing the old school in the 1950s, it reopened for a time to alleviate overcrowding. By 1966, the building was gone, a public park in its place.

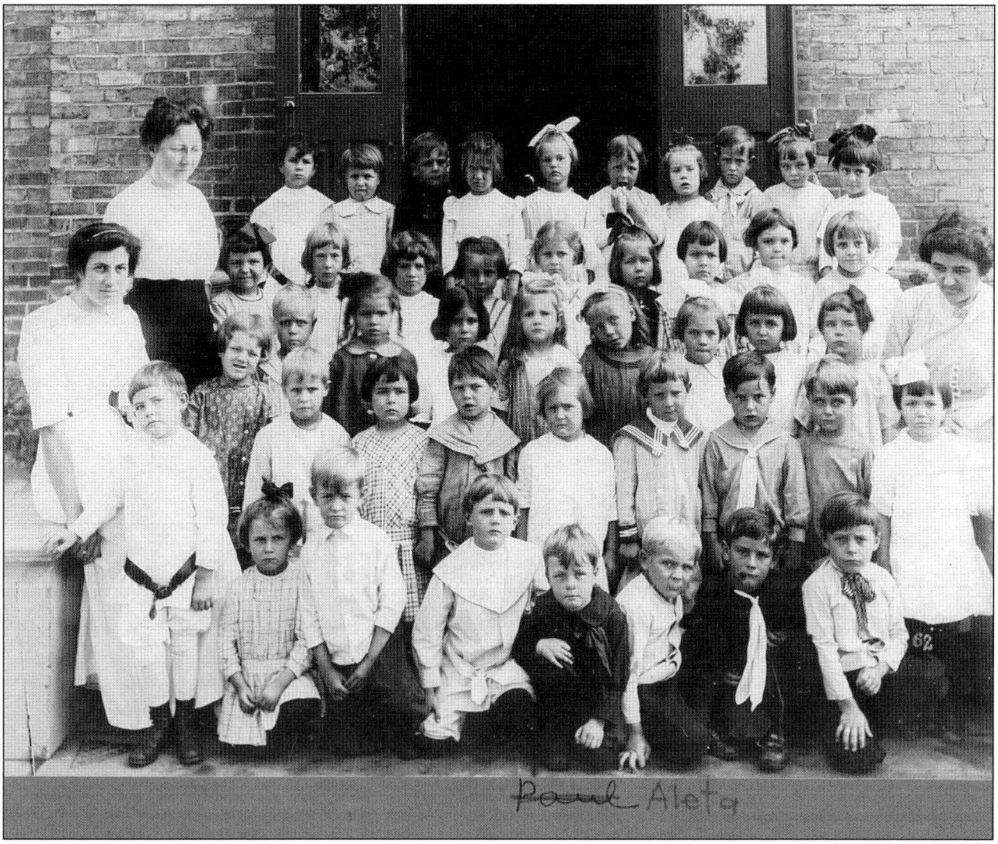

The fourth-grade class at Cherry Street School poses for a photograph on the school steps in 1913. Aleta Ostrander Styers is a student, fourth row, fifth from left. Aleta attended Niles schools and completed teacher training at Western Michigan University. Styers worked for the local school system for many years, teaching at Eastside Elementary from 1943 until her retirement in 1970.

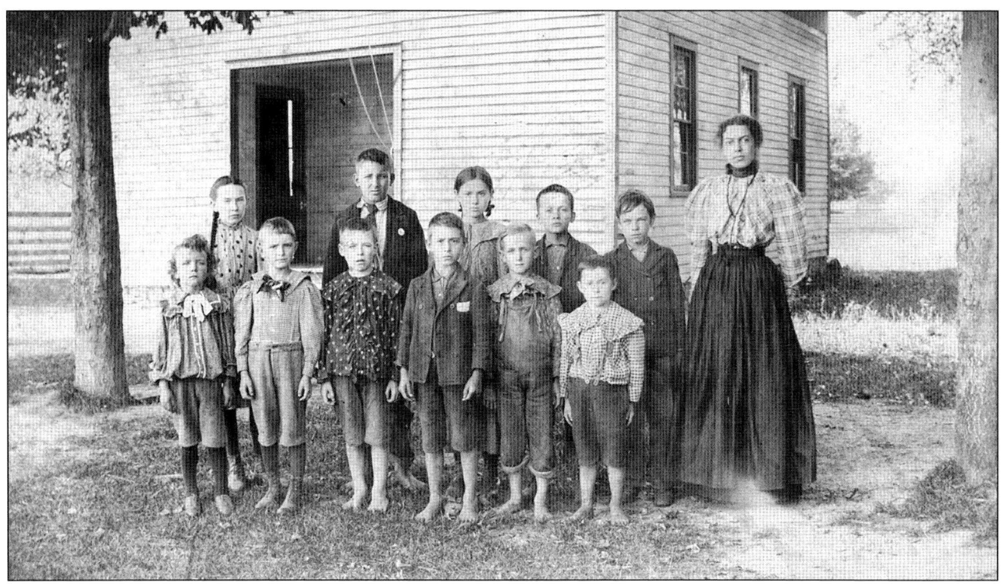

The Pucker Street School was in a rural area just north of Niles city limits. The one-room schoolhouse educated students of various ages. The school operated from the 1870s until it closed in 1963. The building was expanded over time. Like many schoolhouses of the time, it was also used for community meetings. This teacher poses with her class of 11. Several of the children are barefoot. By 1909, as seen in the photograph below, the class has grown, and the children are dressed for picture day.

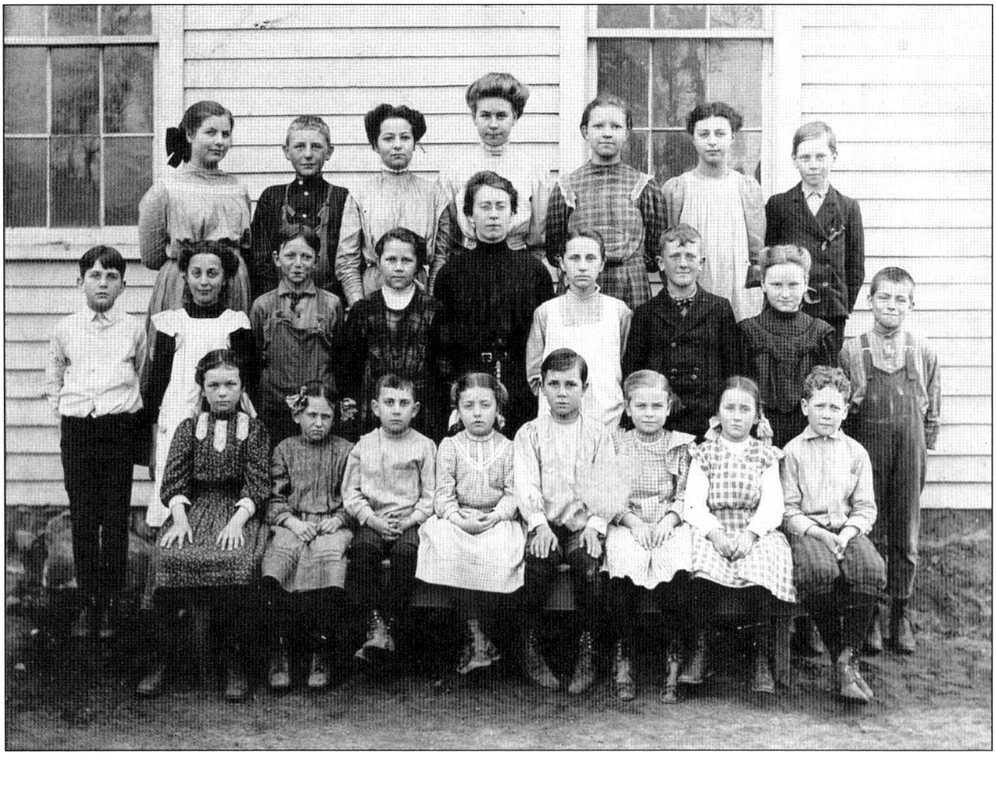

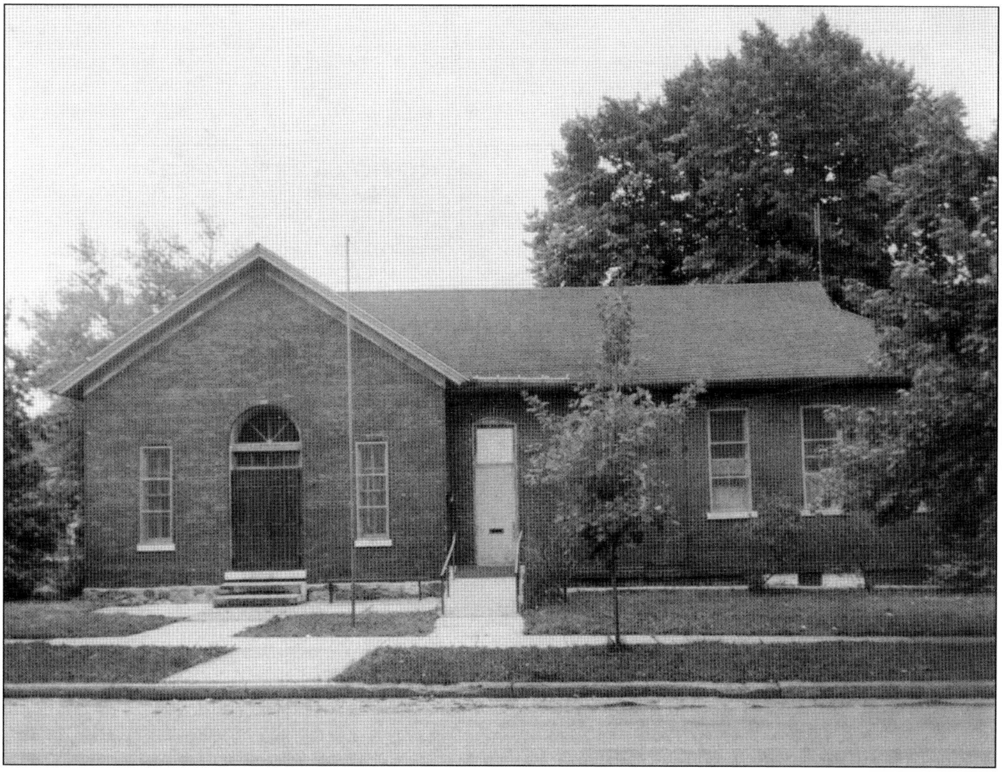

A "colored" school opened in Niles at the corner of Seventh and Ferry Streets in 1867. Schools were integrated a few years later, and the school served all the children in the Ferry Street neighborhood, as seen in this 1906 class photograph. Originally a one-room schoolhouse, a west wing was added in 1903. The building became a Michigan State Historic Site in the 1970s and, from 2005 to 2023, housed the Ferry Street Resource Center. Tragically, the school suffered a fire in 2023. The front façade still stands as a monument to Niles history.

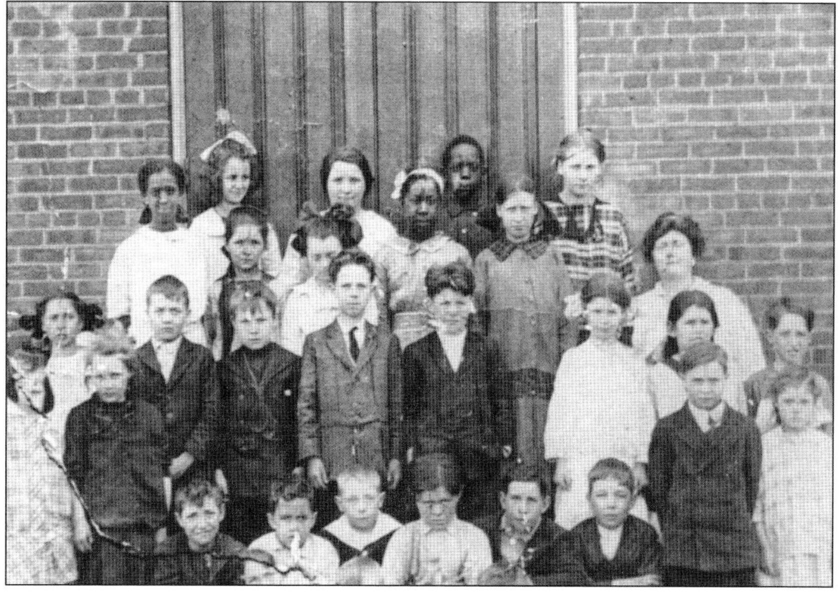

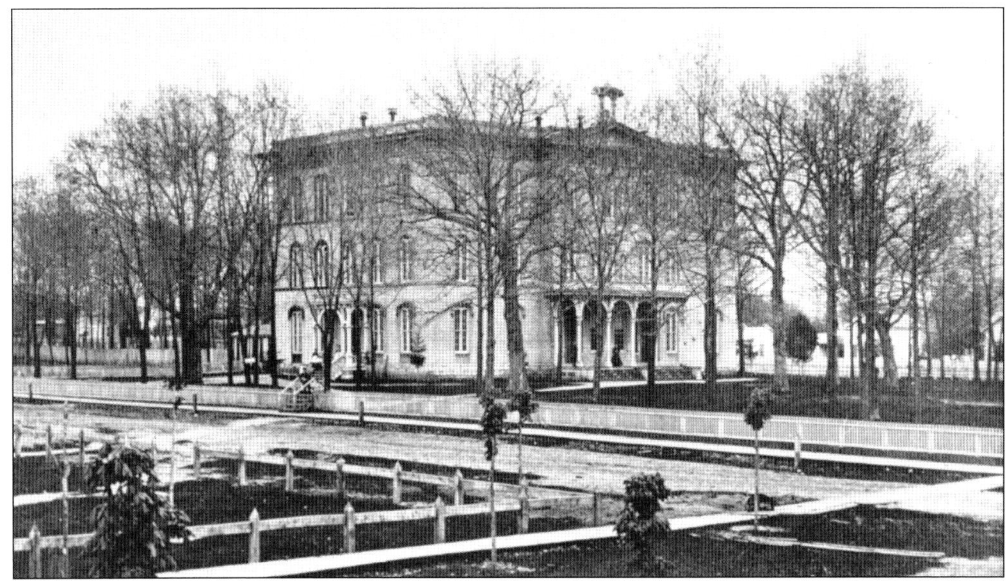

Union School was constructed at Broadway and Seventh Streets. The first classes were held on Monday, September 29, 1856, with 450 enrolled in mainly primary grades, and some high school students from Niles, Sumnerville, Eau Claire, Edwardsburg, St. Joseph, New Buffalo, Elkhart, and Goshen. Out-of-district students paid a tuition fee. School records show coed classrooms each holding 40 to 50 students. Union School was sometimes referred to as "the Central School."

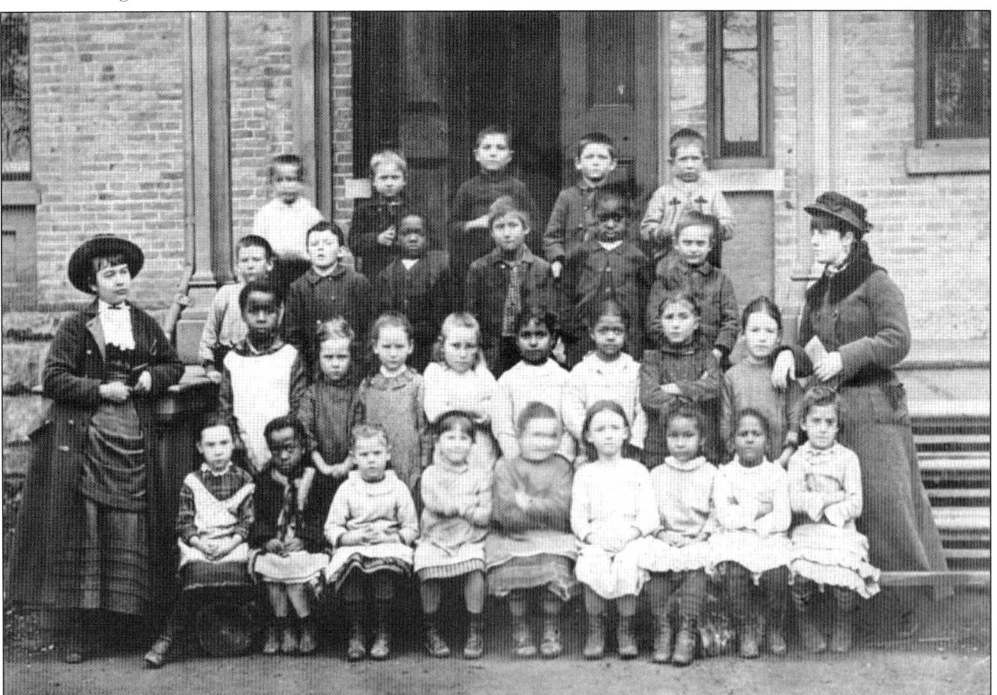

This photograph shows an 1880s elementary class at the Central (Union) School. Belle Finley and Mary Woodruff are identified as the teachers. Both Finley and Woodruff graduated from Niles Central High School in 1881 and took jobs as teachers. Students at Central would have come from all four wards of Niles.

An eighth-grade class from Niles Central (Union) School, room 11, poses in 1888. Sadie Wheaton was the teacher, and J.D. Schiller served as superintendent of the district. Wheaton was paid a salary of $35 per month. Several of the students are members of early Niles families, including Helen Woodruff, Mary Huston, and Guy LaPierre.

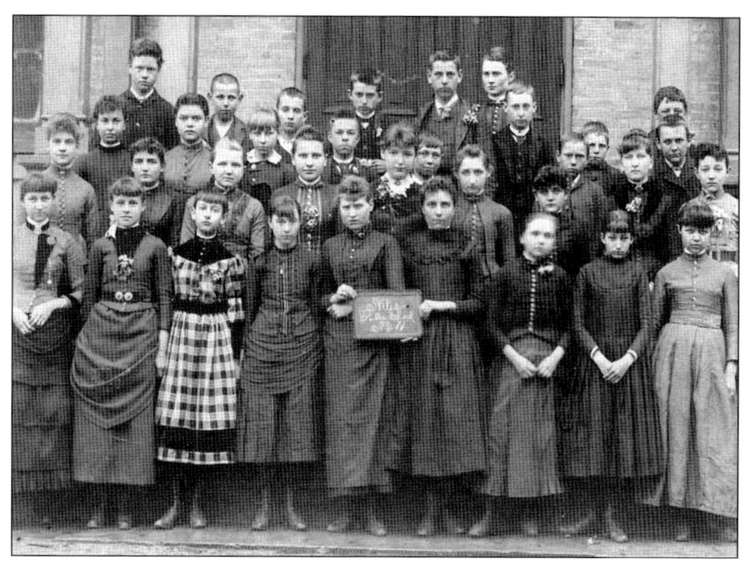

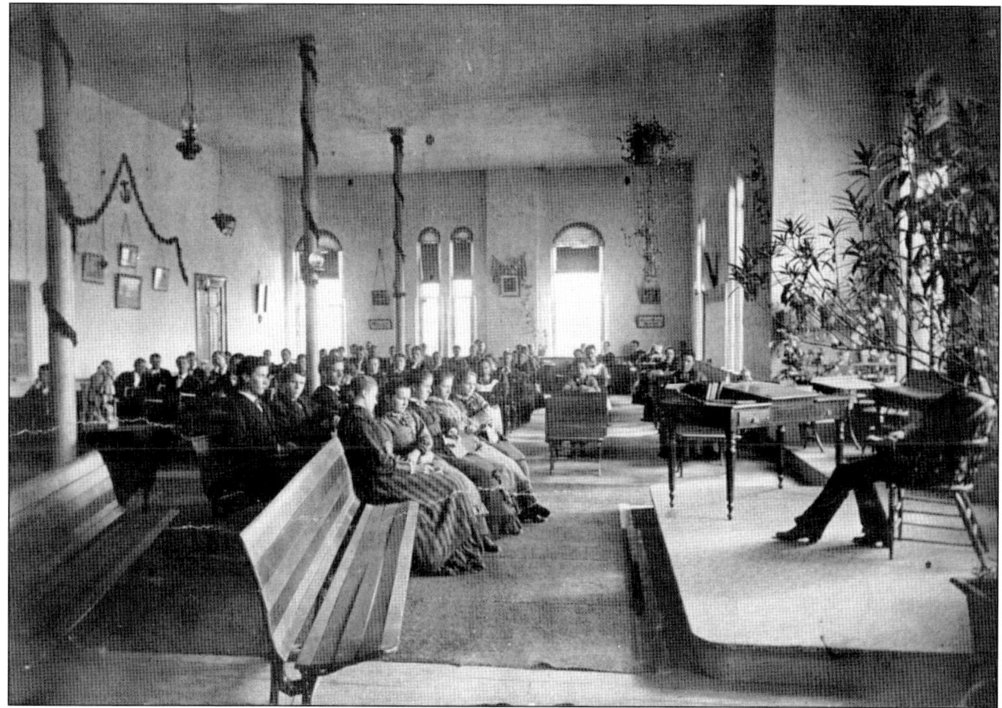

Niles High School students pose for a photograph that was displayed at the Philadelphia Centennial Exposition. Schools from across the country sent photographs to the event held in celebration of America's 100-year anniversary in 1876. The photograph was taken at the Union School. The melodeon at the front of the room is part of the Fort St. Joseph Museum's collection.

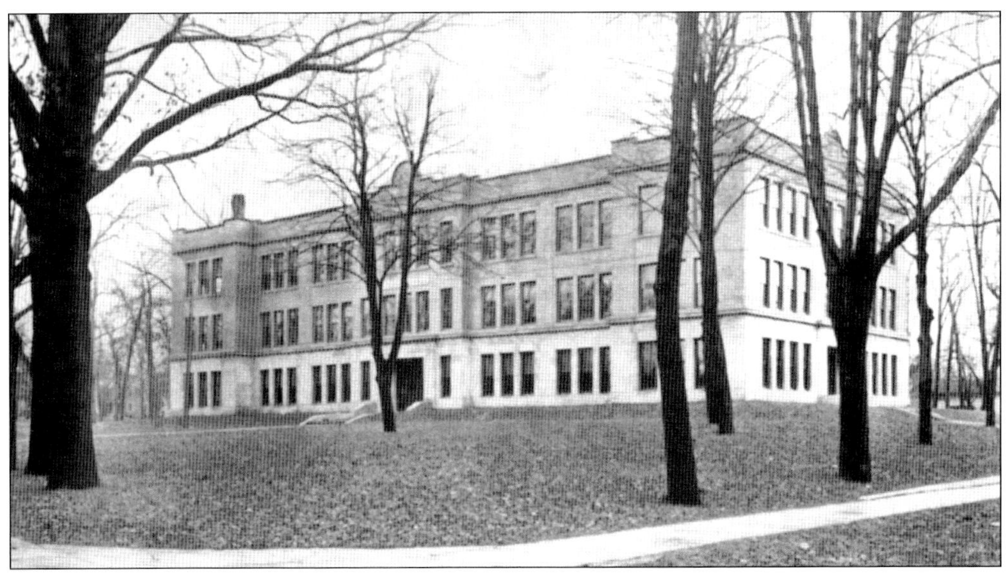

A new Niles Central School, pictured above, was built in 1912 next to the Union School. It mainly served as a high school. A second Central School, pictured below, opened in 1923 on the site of the old Union School after that building was razed. The 1923 building served both elementary and high school students. This new building was needed to meet the rapidly growing population due to the influx of railroad employees with families after the classification yards opened in Niles.

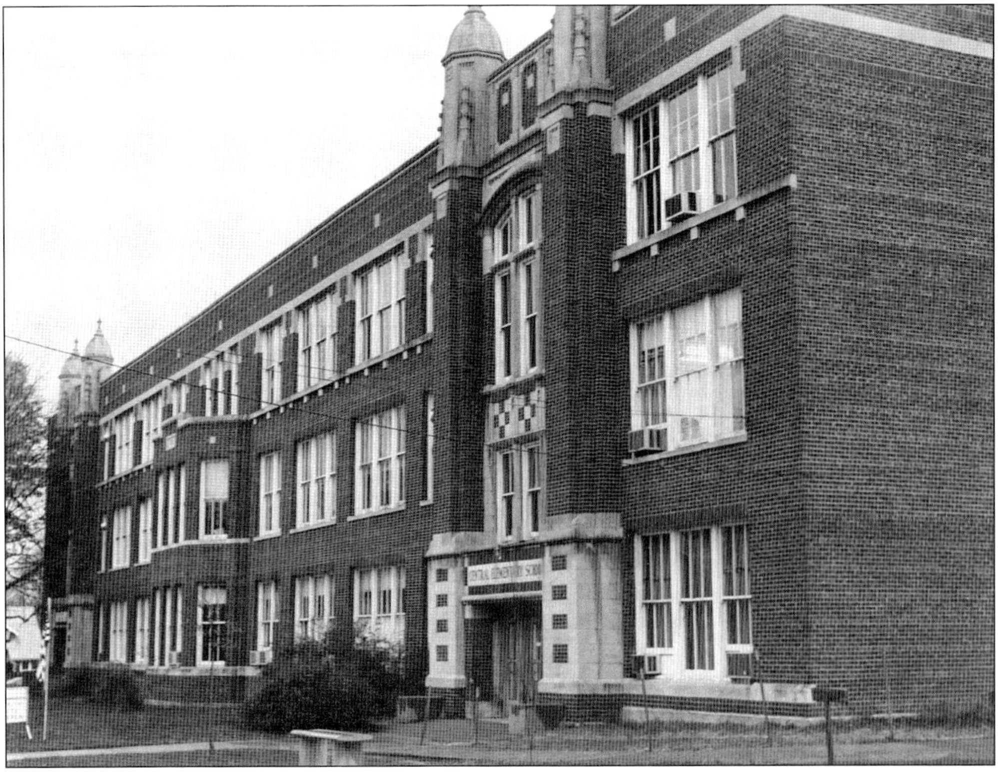

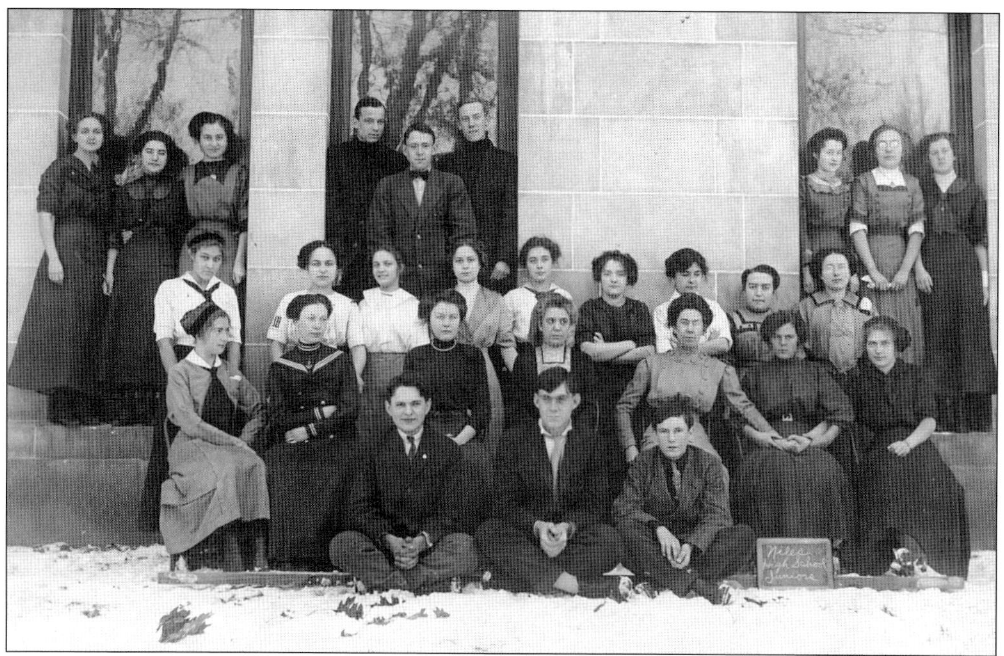

Members of the Niles High School class of 1913 take a group photograph outside the school in December 1911. Hilah Allen, a long-time history teacher at the high school, is seated fourth from the left in the second row. Allen was apparently a popular teacher in the Niles School System. The class of 1919 dedicated the *Tattler* yearbook to her that year.

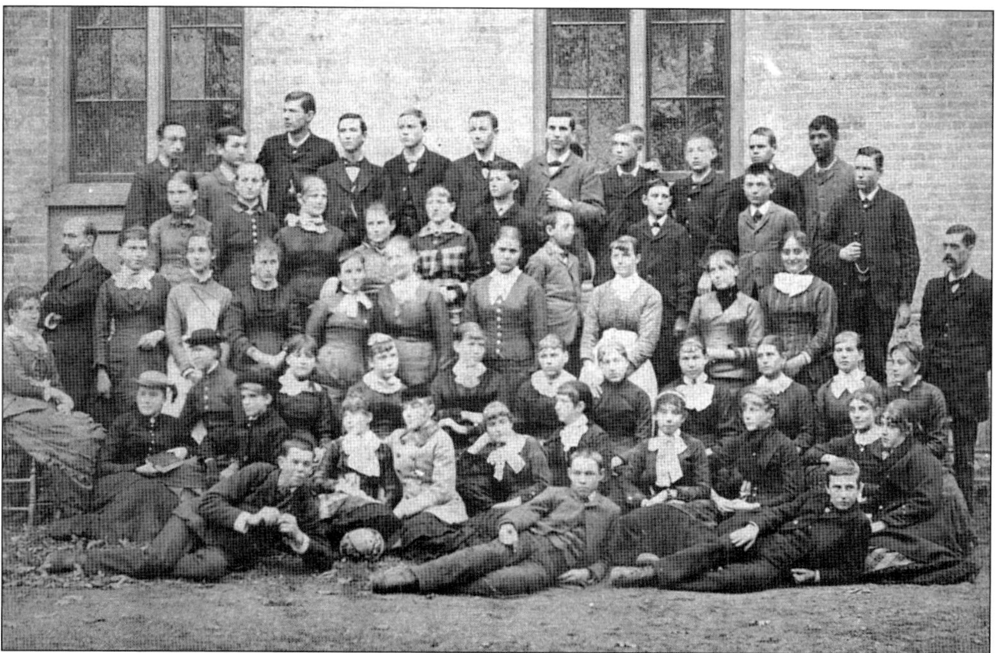

Famous alumni from Niles High School include Dr. Fred Bonine and John Dodge, one of the brothers who founded the Dodge Motor Car Company. Fred and John pose in a somewhat candid shot with the class of 1882 outside the Union School building. They are standing in the back row, with Fred, the tall student, on the left and John, two students to the right of Fred.

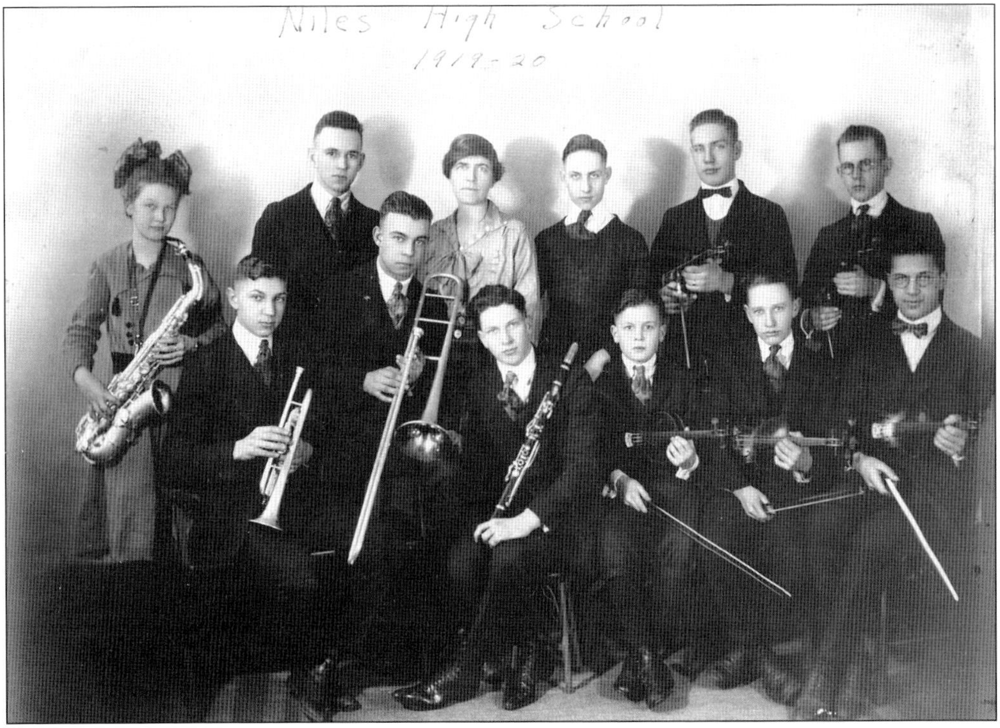

Members of the Niles High School Orchestra for the 1919–1920 school year pose with their director, Lena Lardner (second row, third from left). According to the 1920 *Tattler* yearbook, the orchestra played for many school events, and the group "proved quite indispensable and their music has met with appreciation." Lena Lardner, the sister of author Ring Lardner, directed the Niles High School Orchestra and taught music at the school.

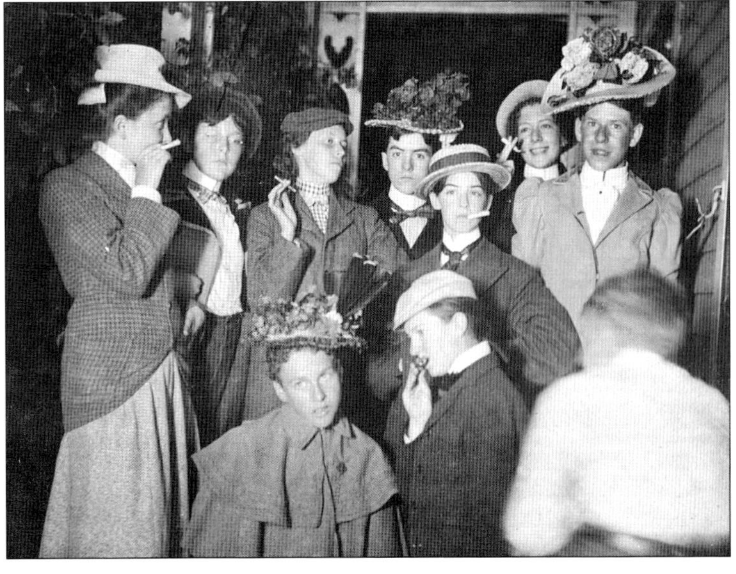

Ring Lardner, second from the left, has some fun with classmates. One of the classmates in the photograph is Orrill Coolidge. Both Lardner and Coolidge graduated from Niles High School in 1901. Lardner became a famous author and Coolidge a librarian. This photograph was taken at the Beeson residence by Marshall Walker, but why they are cross-dressed is a mystery.

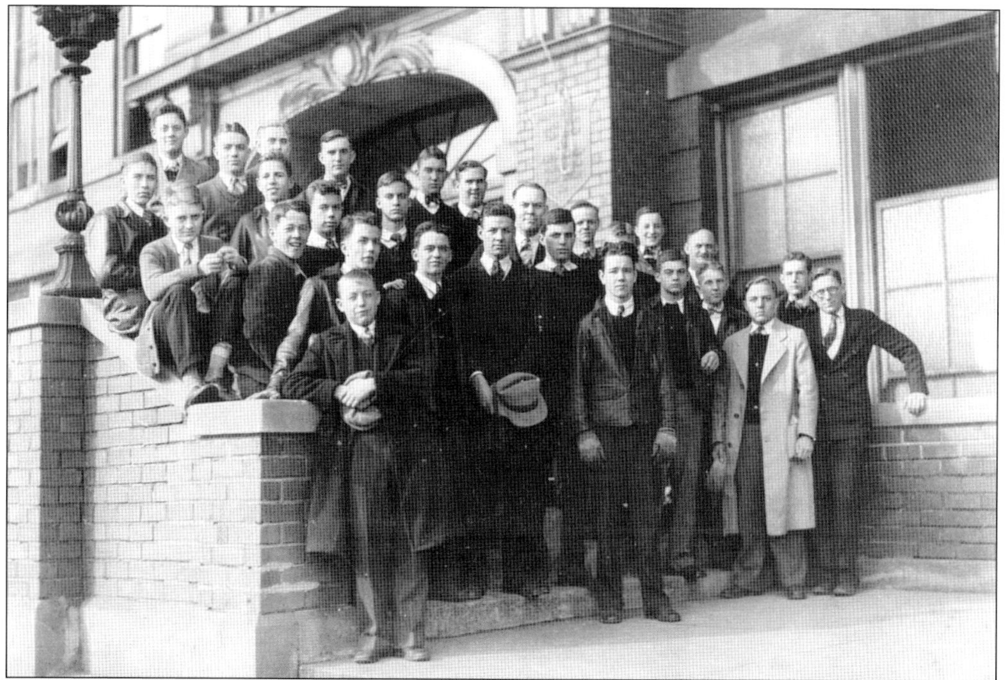

In one of the many extracurricular organizations offered at Niles High School in the 1920s, members of the Hi-Y Club gather outside the school in 1928. Affiliated with the YMCA, Hi-Y clubs for boys were formed at middle and high schools throughout the country in the early 1900s. Members participated in community service projects.

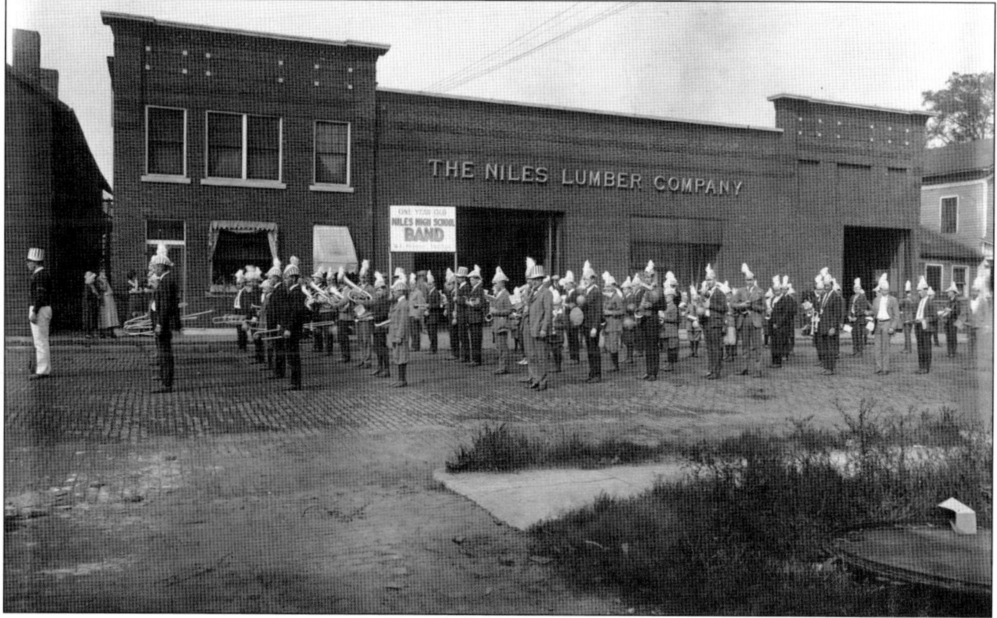

The Niles High School band prepares for a parade near the Niles Lumber Company on Front Street. The sign in the photograph notes the band is one year old, dating the photograph to 1923. William E. Mathews directed the band with over 95 men playing that first year. As many had never played an instrument, Mathews trained many of the musicians himself.

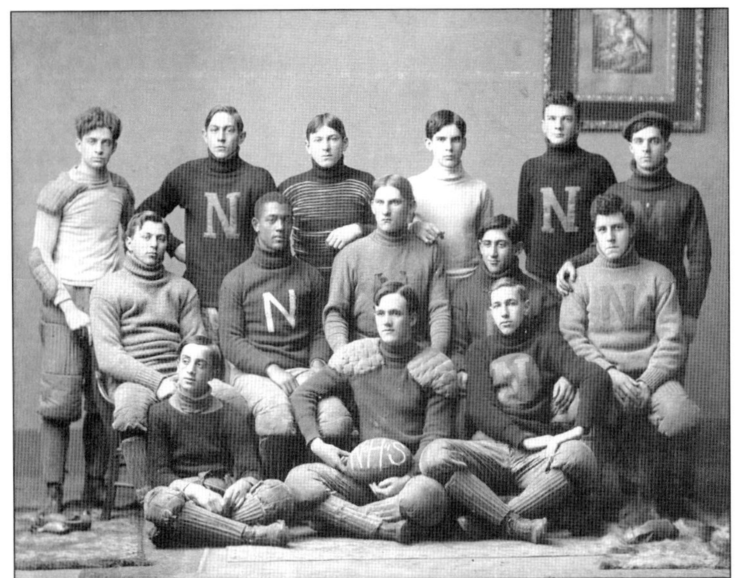

The 1904 Niles football team is pictured here wearing school sweaters and protective gear. They played five games, with a record of 4-1. On Thanksgiving Day, they played the 1894 alumni team and defeated the alums, 50-0. However, the 1894 team ended up becoming the Michigan Amateur Football Co-State Champions that year, along with Kalamazoo.

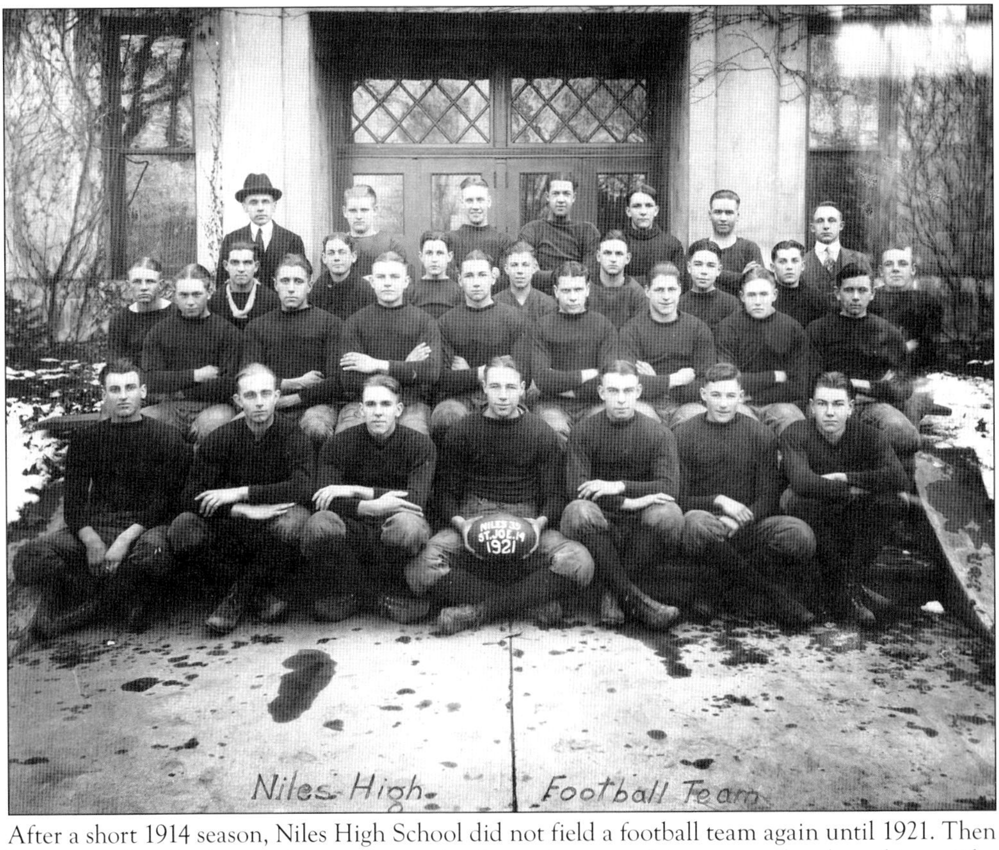

After a short 1914 season, Niles High School did not field a football team again until 1921. Then known as the Cardinals, the team wore leather helmets, red jerseys, and red stockings. The Cardinals defeated St. Joseph High to become the Berrien County champs. Dowagiac handed the team their only loss of the season. The rivals met in Dowagiac on October 26 with the final score of Niles 6, Dowagiac 13.

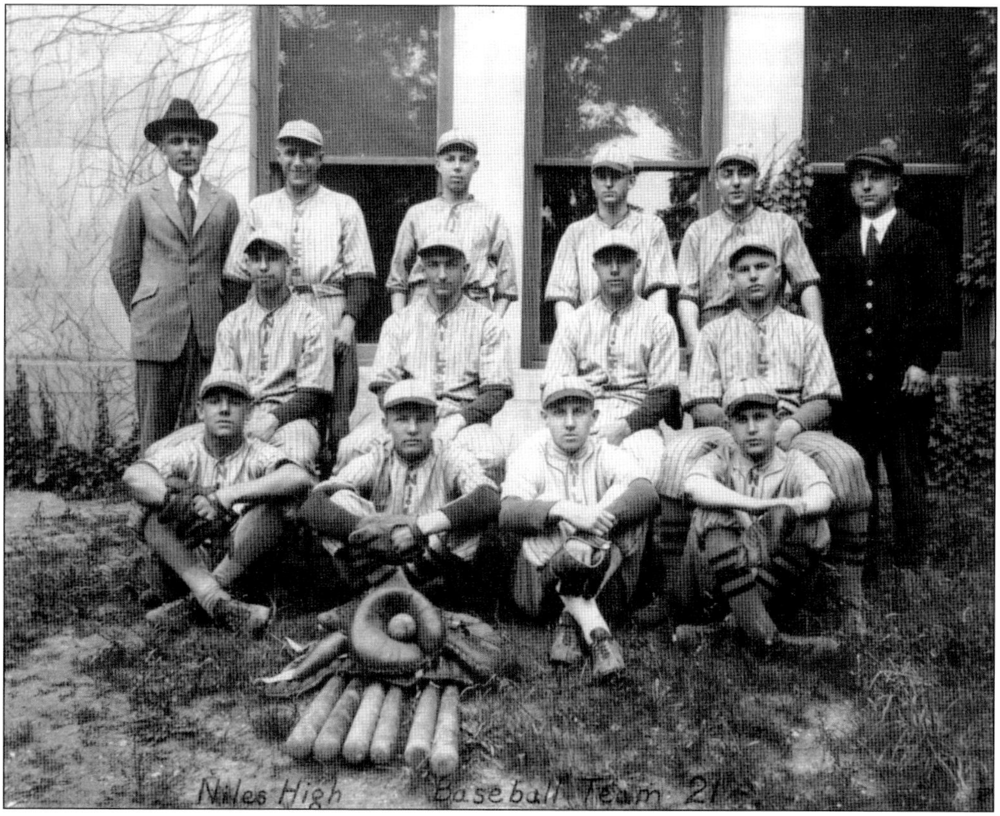

The Niles baseball team of 1921 was plagued by injuries but still managed to win five of their eight games. Rival school St. Joseph bested Niles for the county championship. L.S. Walker (third row, far right) was the team coach, and Walter Zabel (third row, far left) was the manager. Third baseman Oliver Lee (second row, second from left) was the team captain.

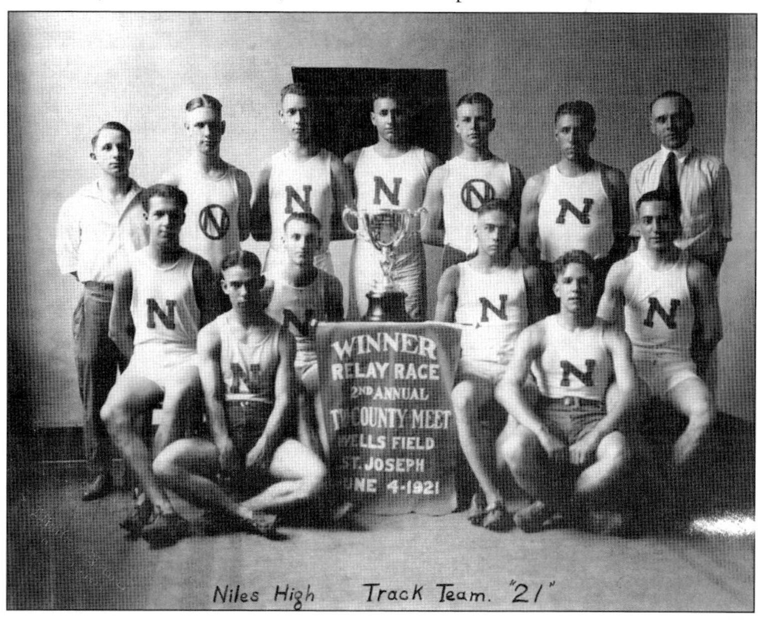

Niles found success with their 1921 track team. On June 4, Niles took top honors at a Tri-County meet held at Wells Field in St. Joseph. Niles won the final relay race, despite protests from St. Joseph that one of their men was tripped by a Niles runner. The judges ruled the race fair, and Niles took home the silver trophy and the banner seen in the photograph.

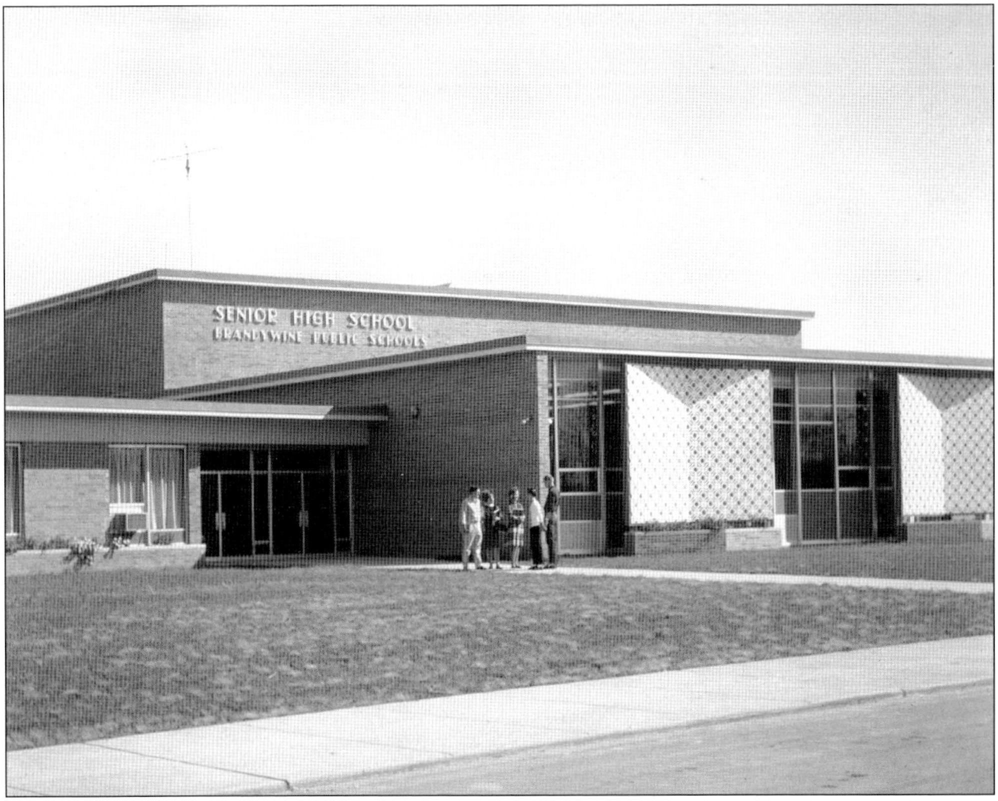

Located in South Niles Township, Brandywine School District built this senior high school in 1961. According to a *Niles Daily Star* article from December 16, 1961, the new school building could accommodate up to 1,200 students and included fully equipped classrooms for subjects like language-learning and home economics, along with a large combination gym/auditorium.

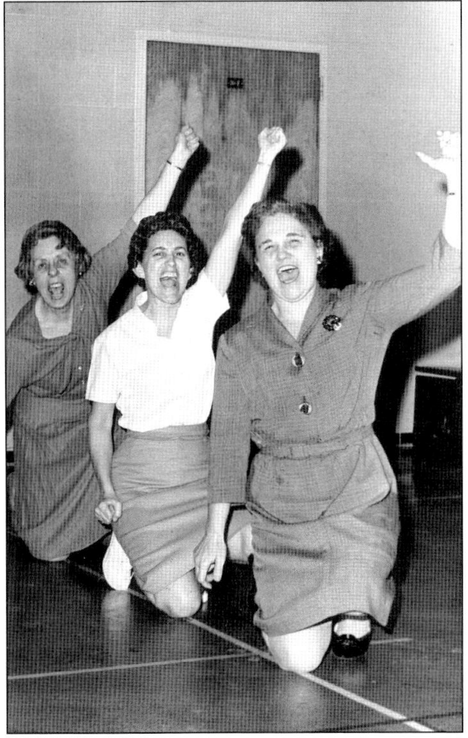

From left to right, Margaret Parrott, Lemoine Bogue, and Lillian Luckey practice for the 1966 Niles High School faculty-senior class basketball game. Several school staff members were cheerleaders for the benefit game, which helped raise money for school improvements. Lillian Luckey, a 1937 graduate of Niles High School, worked as a secretary for Niles Community Schools from 1959 to 1984. Parrott and Bogue were faculty members.

Seven

Arts, Leisure, and Recreation

As Niles grew, so did opportunities for entertainment, community engagement, and recreation. The upper floors of Main Street buildings often included halls for meetings, social events, and performances. By the mid-19th century, Niles was a regular stop for traveling troupes. Among them were the Peak Family Swiss Bell Ringers, who performed nationally and made Niles their home. In 1868, Peak Hall, named after the famous bell ringers, opened on Main Street. Later known as the Niles Opera House, the grand venue hosted concerts, plays, operas, vaudeville shows, and graduation ceremonies. With the rise of movie theaters in the early 20th century, its use declined. The Riviera, Ready, and an outdoor theater on the edge of town offered modern entertainment.

Fraternal organizations like the Elks and the Masons raised money for civic projects. Barred from joining these clubs, African Americans formed their own groups. Harrison Lodge No. 6, Michigan's first African American Masonic lodge, started in 1859. At a time when women were expected to remain in the domestic sphere, clubs offered empowering opportunities for learning, activism, and community. Reading clubs and sewing circles thrived. Other women's groups focused on service. The Women's Progressive League led projects to install historic markers, establish Island Park, Niles's first public park, and register women to vote after the passage of the 19th Amendment.

Churches played an early role in cultural life for many, not only as places of worship but also as centers for fellowship. Clubs and organizations were as much a part of the community as business and industry. These groups met in members' homes, local churches, or places like the Pickwick Club.

Niles's waterways provide leisure activities. Barron Lake once housed a number of destination resorts with beach access, picnic areas, dance halls, and live entertainment. Boating and fishing remain popular pastimes on the St. Joseph River.

In the early 1900s, traveling carnivals set up on Main Street, bringing entertainment acts and rides. The Four Flags Apple Festival, Hunter Ice Festival, Riverfest, and the Fort St. Joseph Archaeological Open House continue to draw visitors and celebrate history and culture in Niles.

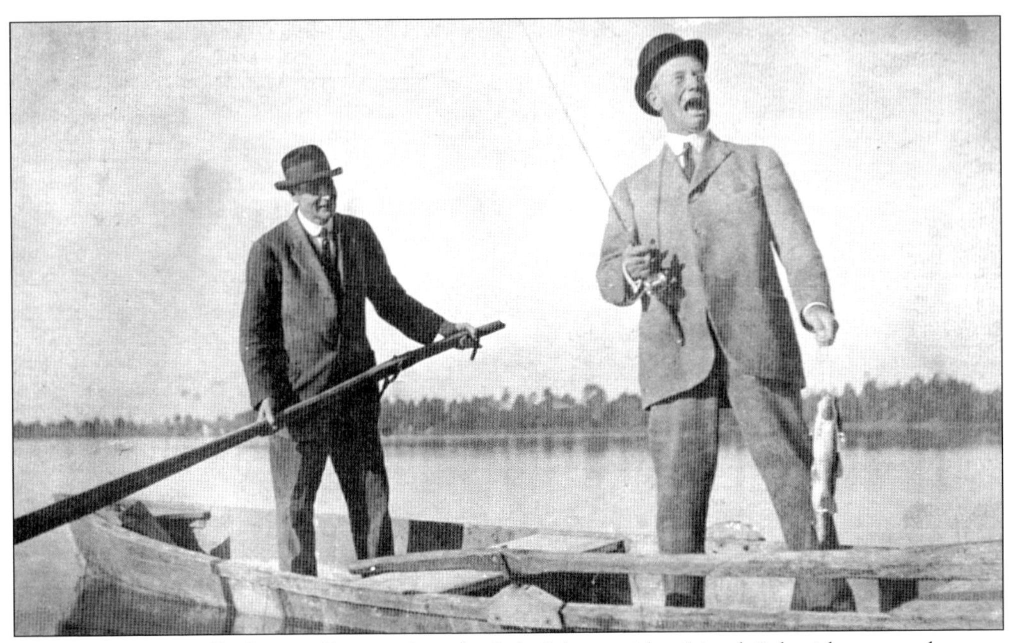

F.C. Schmidt gleefully shows off a prized catch while fishing with a friend. Schmidt operated a meat market on Main Street that opened in 1864. Notice the men's business suit attire. Niles's rivers, creeks, and Barron Lake provided ample fishing opportunities for residents.

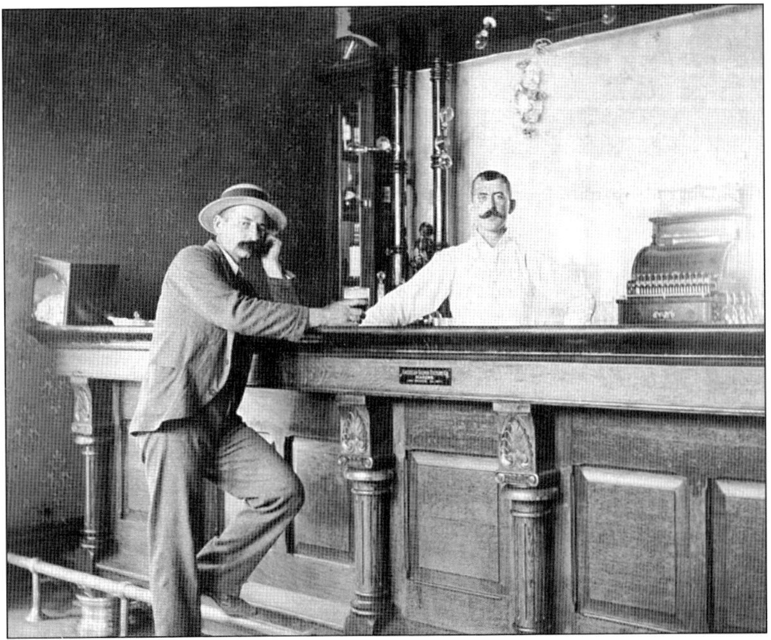

Eben Gage stands behind the bar at a saloon serving a customer who appears to need it! Gage eventually moved to the South Bend area and worked "at the helm" of other bars that advertised his services in an attempt to woo Nilesites. Gage did make at least one mistake. In 1907, he pleaded guilty to serving a minor and paid a $30 fine.

William Peak organized the Peak Family Swiss Bell Ringers in Medford, Massachusetts, in 1838. The musically talented family played several instruments. The Peaks played as many as 42 bells, varying in size from a large cowbell to a small dinner bell. Son William H. Peak is seen here with his harp. The Peak family toured across the country, and several of its members ended up living in Niles.

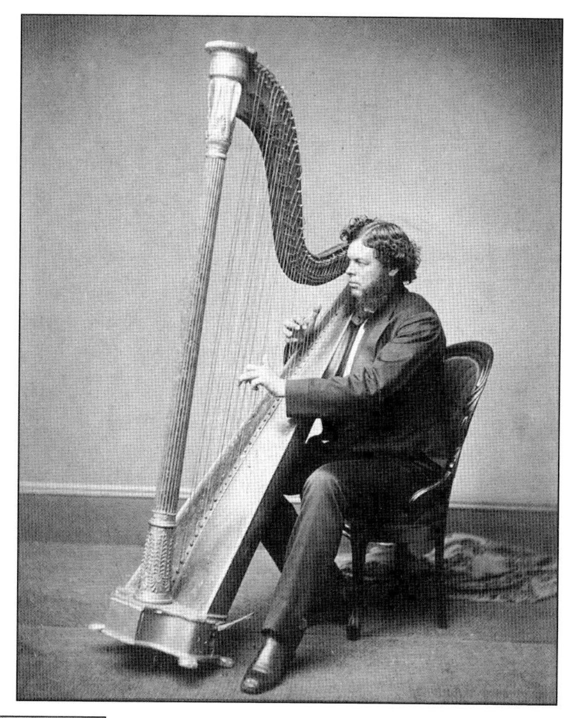

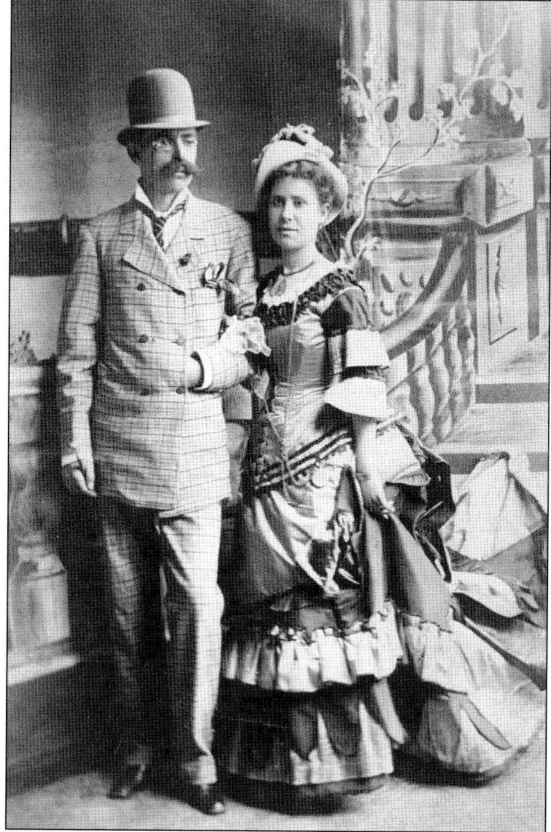

Fannie and Jeppe Delano performed with the Peak family. After the Peak troupe disbanded, Fannie and Jeppe started successful vaudeville careers. Jeppe was primarily a comedian. Fannie played bells and the horn and was known for her legendary soprano voice. The couple married and toured through leading theaters across the United States, Canada, and the West Indies. Upon retirement, they settled at their Niles home.

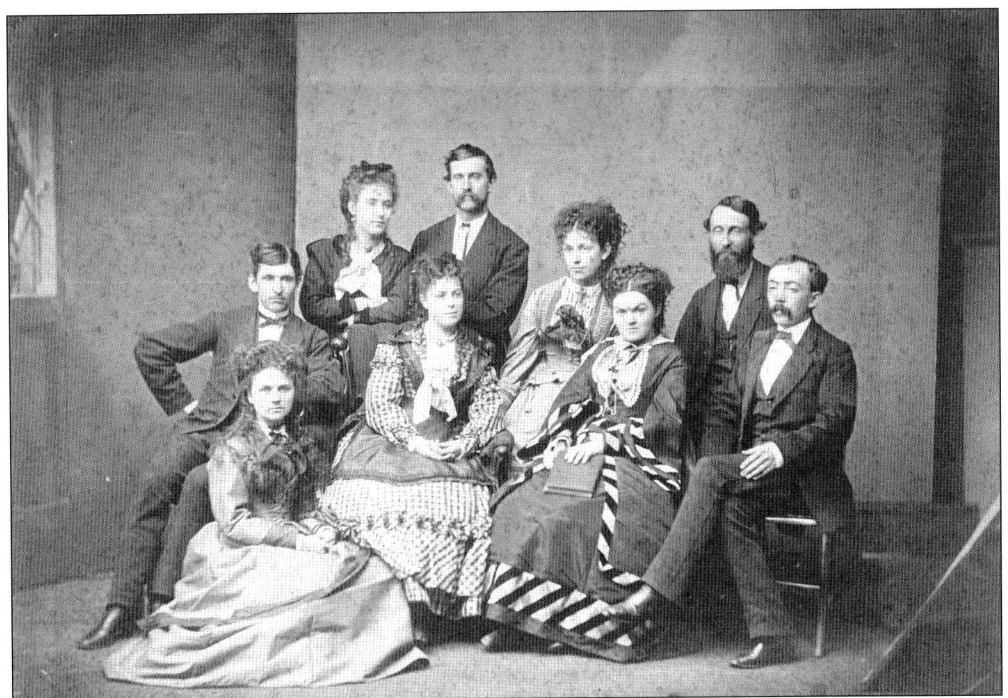

The Berger family performed with the Peak family during the 1860s. Together, they toured the East Coast as a "very superior western troupe," complete with fancy costumes and music of the modern era. The families performed their last show in 1869 when many of the Peaks retired. The Berger family moved to Jackson and continued touring, returning often to Niles, where they still had a loyal following.

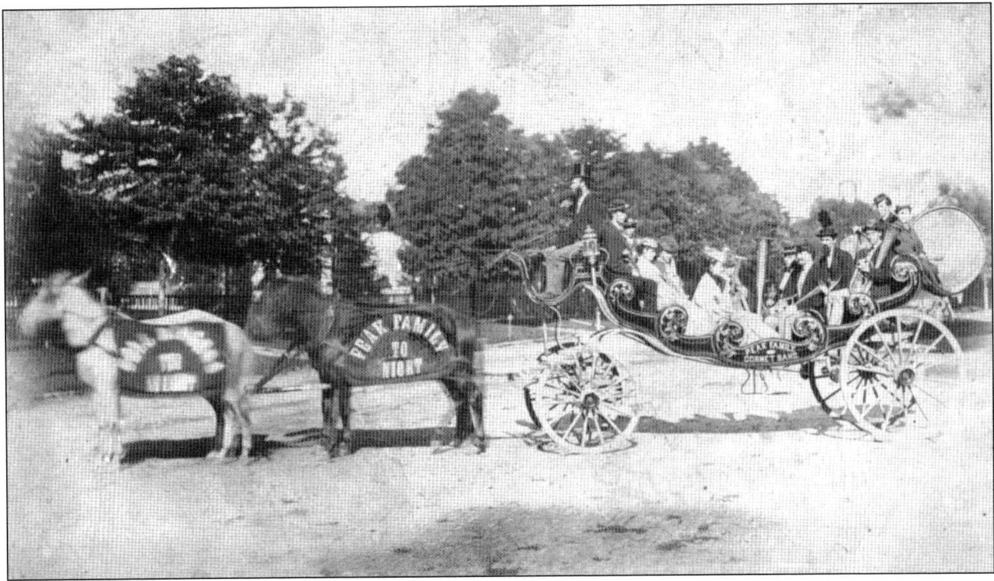

Niles mechanic E. Murray built this traveling wagon for the Peak family in the 1860s. It featured a unique design for touring musicians, described as jet black with fine ornamental painting. Banners proclaiming, "Bell Ringers Tonight," drape the horses. Crowds watch as the wagon passes by, filled with musicians and their instruments.

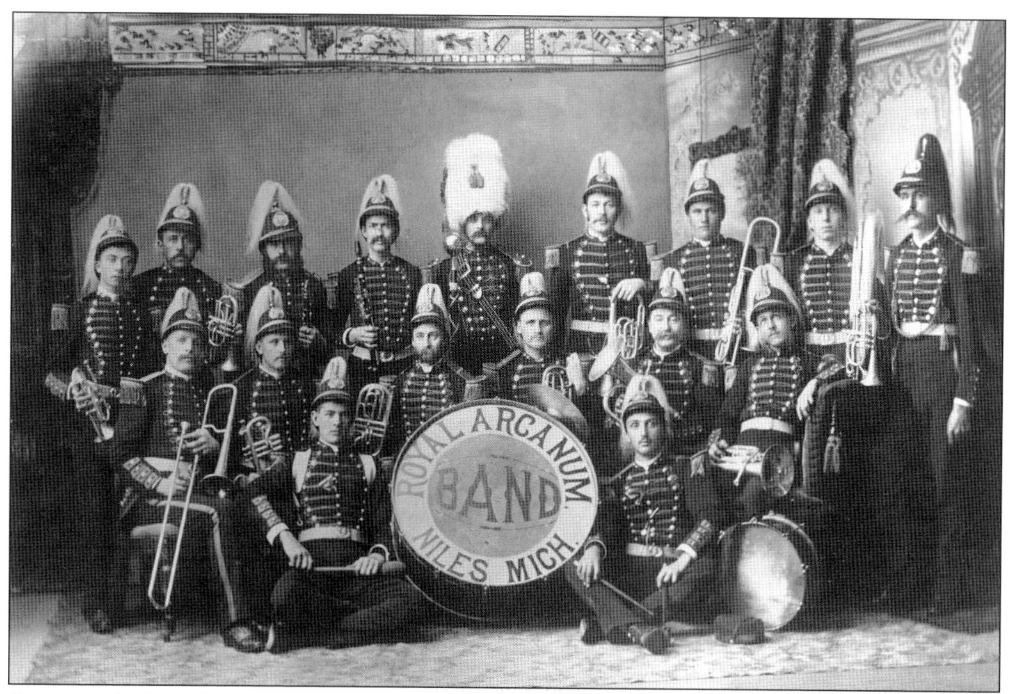

Comprised of members of the fraternal benefit organization of the same name, the Royal Arcanum Band is mentioned in an 1885 news story escorting Civil War general Henry Morrow through Niles. The band played for various civic, political, and patriotic events, including a memorial held in Niles to honor Pres. Ulysses Grant.

The Niles Citizens Band was formed in 1900 and featured two female musicians. The band played at street fairs and events in Michigan City, St. Joseph, Bay City, and of course, Niles. Cornet player Zola Davis pursued a musical career, joining the Waterloo Ladies Military Band and traveling across the country.

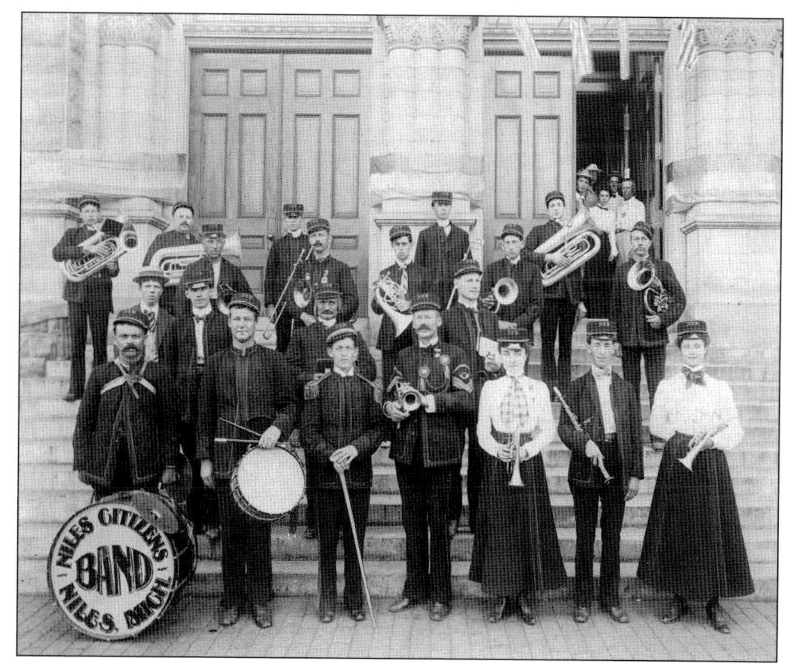

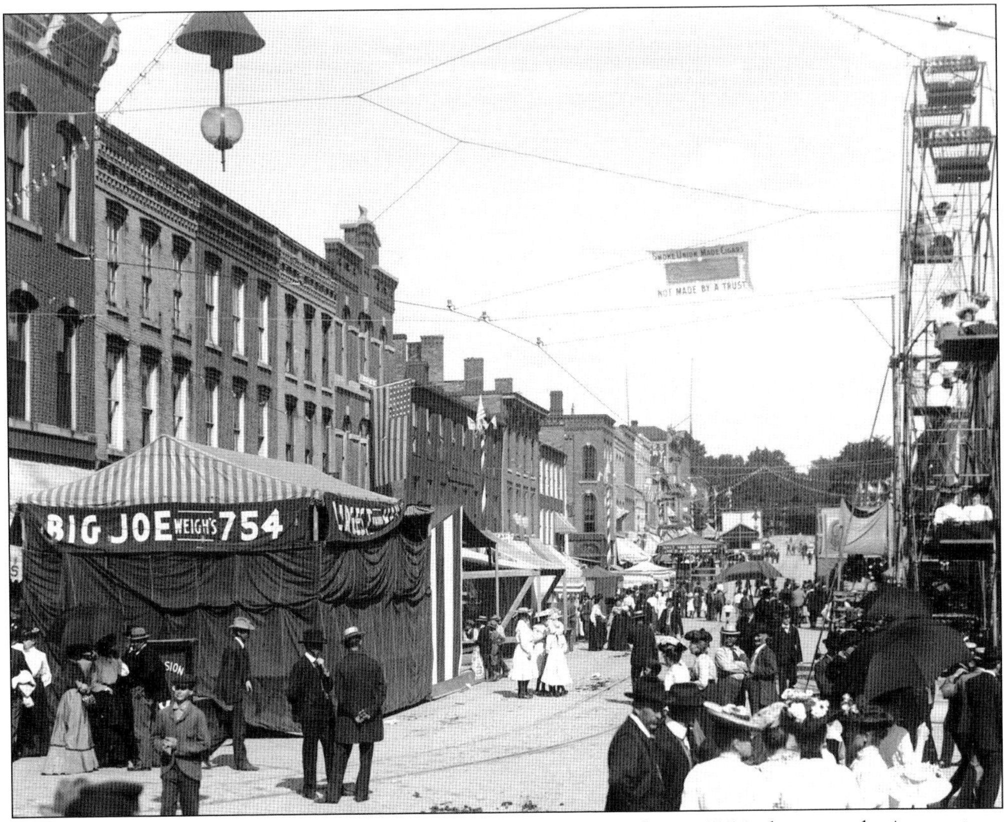

Traveling carnivals regularly set up on Main Street, as seen in this c. 1904 photograph. Attractions included a Ferris wheel, a tent with "Big Joe," who weighed 754 pounds, and several other exhibits and rides. The sponsoring sky banner encourages people to "Smoke Union Cigars, not made by a trust." The interurban streetcar track is visible in the road (along with circus animal droppings).

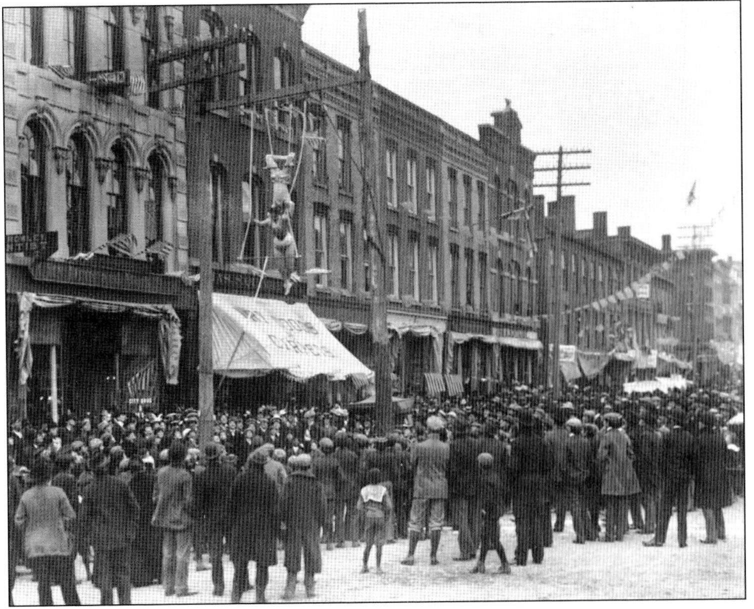

Trapeze artists performing in downtown Niles draw a large crowd of spectators. The traveling act set up on Main Street between Front Street and the 200 block. Rowley Dentist, City Drug and Book Store, and Charles S. Quimby's Dry Goods and Carpets are visible in the background.

A group gathers in front of the Riviera Theater in 1940. John Baumann opened the Riviera on North Second Street in 1920. The theater was notable because it did not have segregated seating, unlike many theaters at the time. The Riviera closed in 1960, and the building was razed in 1970 as part of urban renewal. This photograph also shows the downtown holiday decorations on the streetlight.

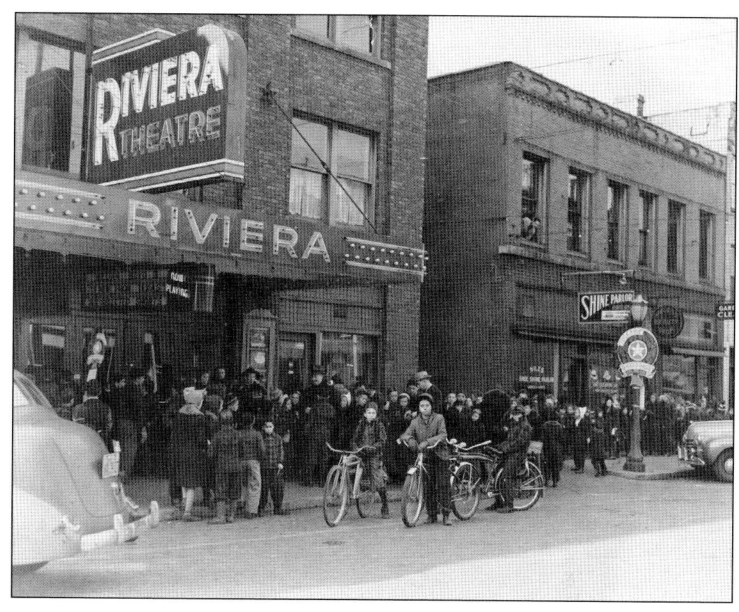

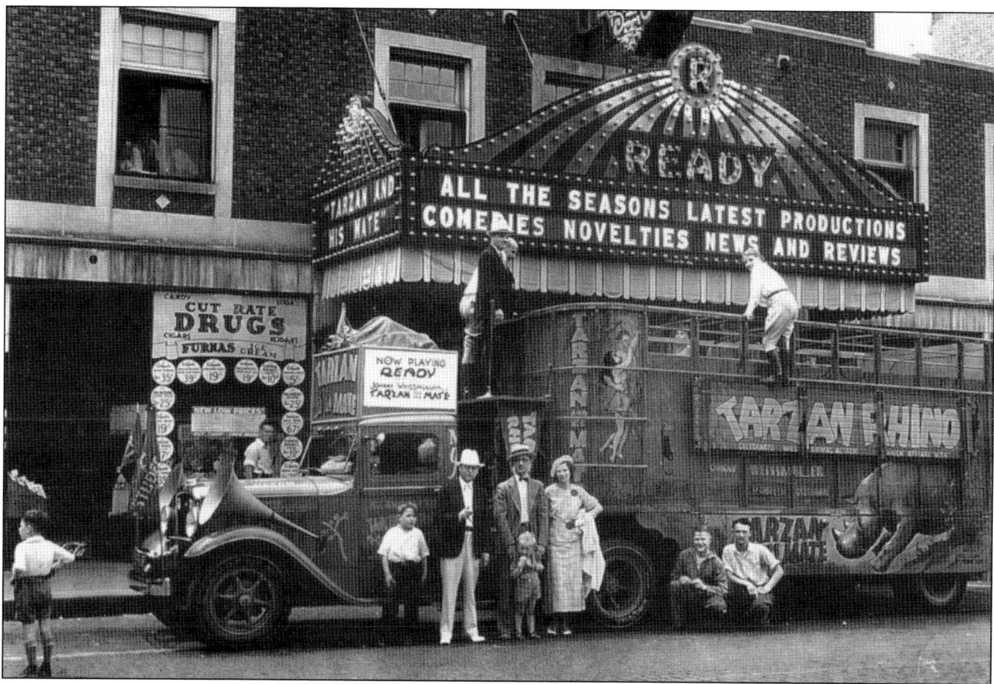

An advertising truck parks in front of the Ready Theater to promote the 1934 film *Tarzan and His Mate*. Built in 1927 by businessman Thomas Willard Ready, the premier local theater was the site of many movie screenings and community events. The Ready featured 1,250 seats and offered a stage and dressing room. In 1928, it was the first theater in town to show a "talkie."

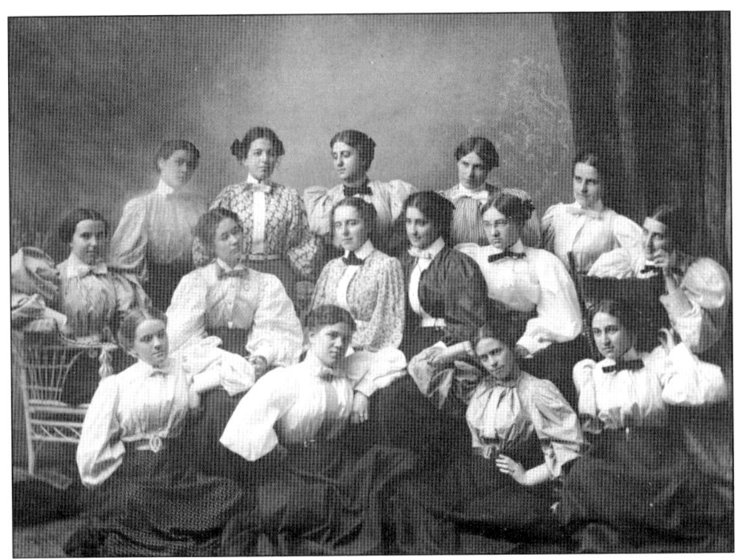

Originally called the JFF (Just for Fun), this long-running women's club later changed its name to the Seepewa Reading Club. Seepewa is a Miami word that refers to a body of water. The club was launched in the 1890s to discuss research topics as well as books. One of Niles's longest-running women's clubs, Seepewa held several multigenerational members.

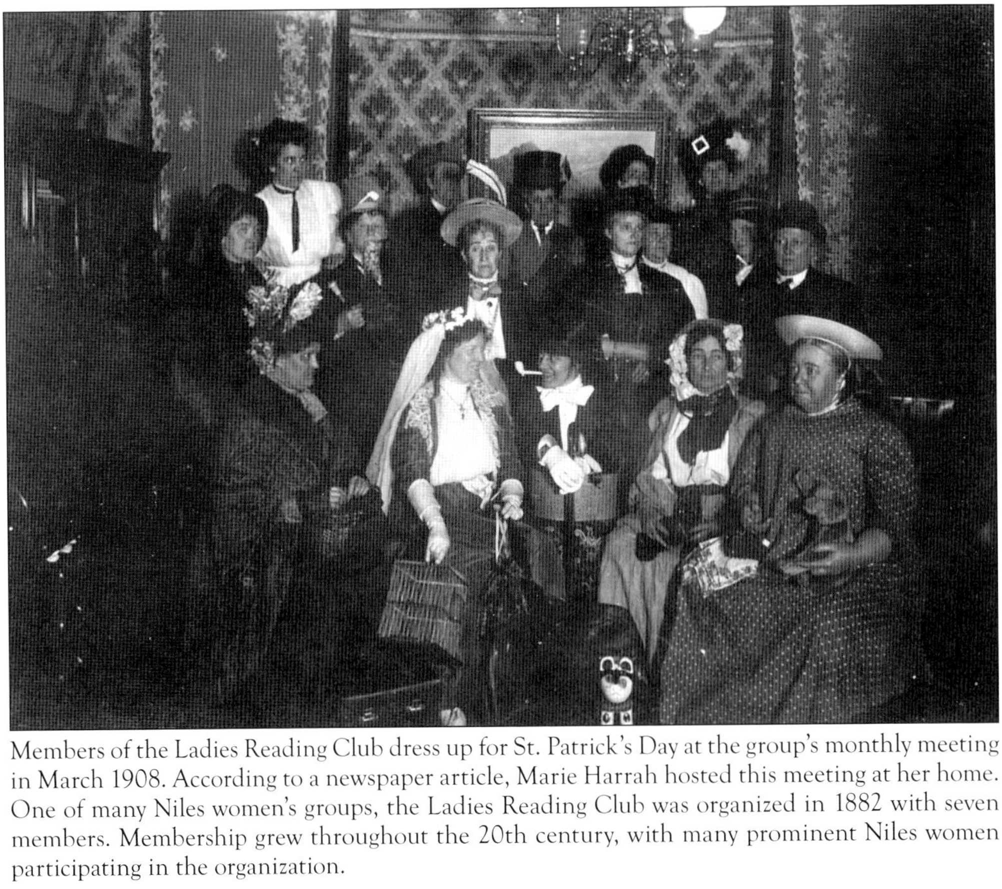

Members of the Ladies Reading Club dress up for St. Patrick's Day at the group's monthly meeting in March 1908. According to a newspaper article, Marie Harrah hosted this meeting at her home. One of many Niles women's groups, the Ladies Reading Club was organized in 1882 with seven members. Membership grew throughout the 20th century, with many prominent Niles women participating in the organization.

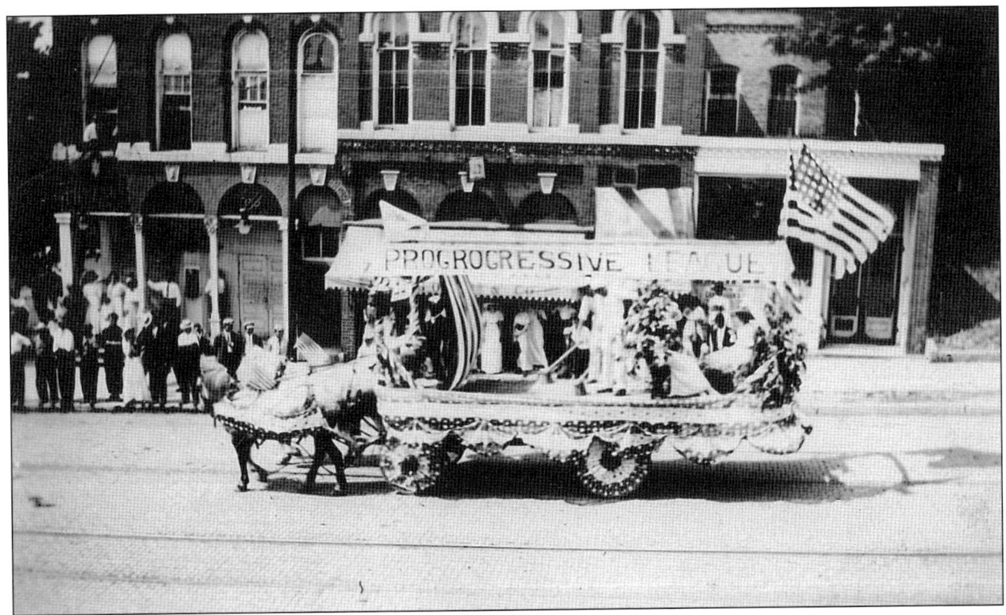

The Women's Progressive League float rolls down Main Street during a July 4, 1913, parade to celebrate Independence Day and the dedication of the Fort St. Joseph Boulder. Formed in 1912, the Women's Progressive League focused on civic improvements, including opening Island Park and placing a stone cross at Father Allouez's gravesite. The league also advocated for the 19th Amendment and helped register women voters in Niles after it passed.

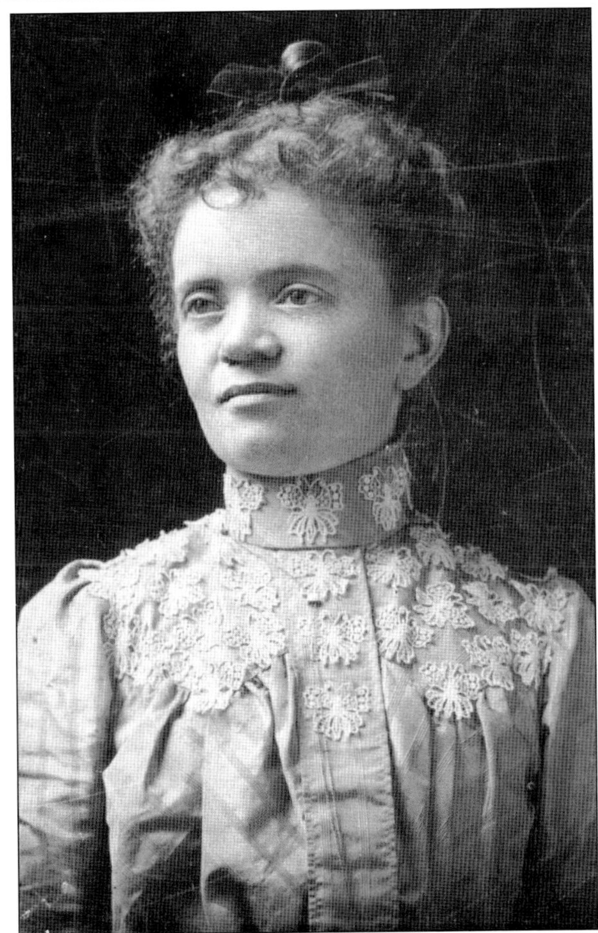

Artist and activist Lottie Wilson's trailblazing career began here in Niles. Born in 1854 to Calvin and Henrietta Hill Wilson, Lottie was the first African American to attend the School of the Art Institute of Chicago. Theodore Roosevelt displayed her painting, "President Lincoln with a Former Slave," at the White House. Wilson actively campaigned for women's suffrage and equal rights for African Americans through prominent national organizations. In 1906, she moved back to her hometown, working as an artist and continuing her advocacy until her death in 1914.

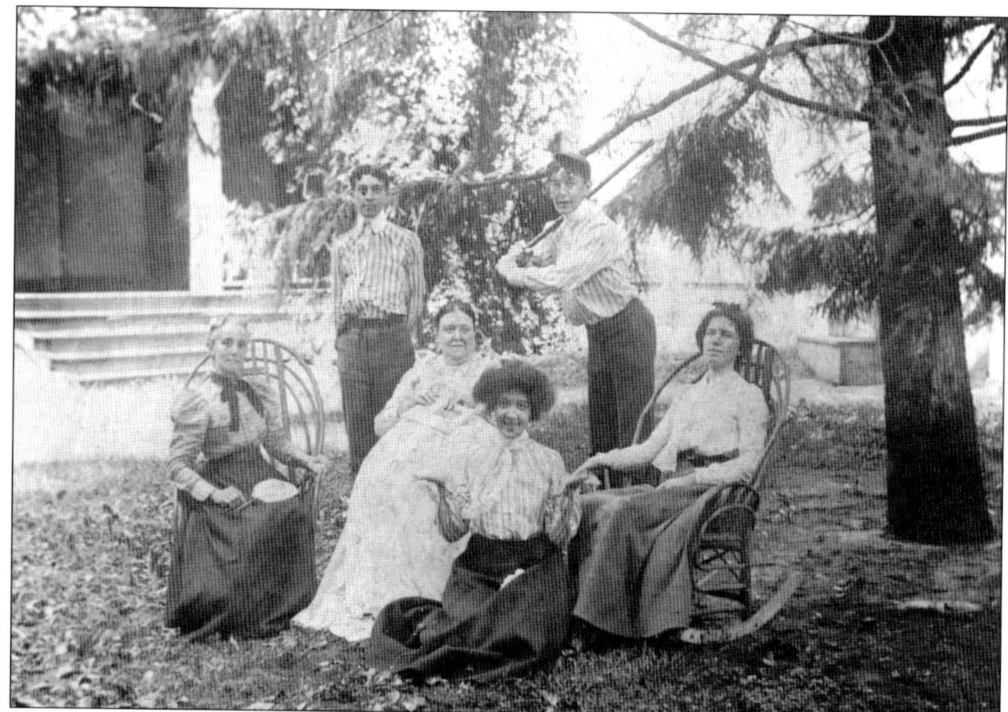

Born in Niles on March 6, 1885, Ringgold Wilmer "Ring" Lardner was a famous author known for his sports reporting, short stories, and satire. The photograph shows a young Ring Lardner, holding a baseball bat, with several members of his family outside their home on Bond Street. The house is still standing.

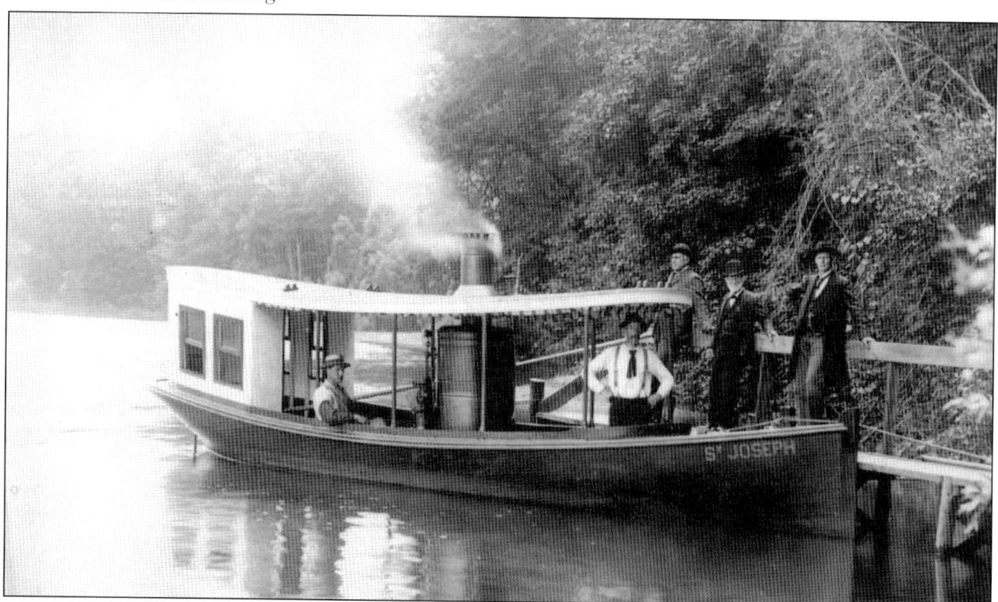

Henry Lardner, father of the famous author Ring Lardner, owned a river steamer called the *St. Joseph*. Lardner often took excursions on the river to South Bend, Indiana, in the late 1800s. Referred to in the newspapers as a "beautiful yacht," the *St. Joseph* appears to have accommodated private parties for the family and also public excursions for a small fee.

The one-man band, Victor J. Hyde, graduated from Niles High School in 1931 and attended the Drake Conservatory of Music. A charismatic showman, Hyde created an act where he would play several instruments at once while twirling a baton and marching. During his career, Hyde played across the country and in Europe but returned home to Niles often, where he participated in local parades and charitable events.

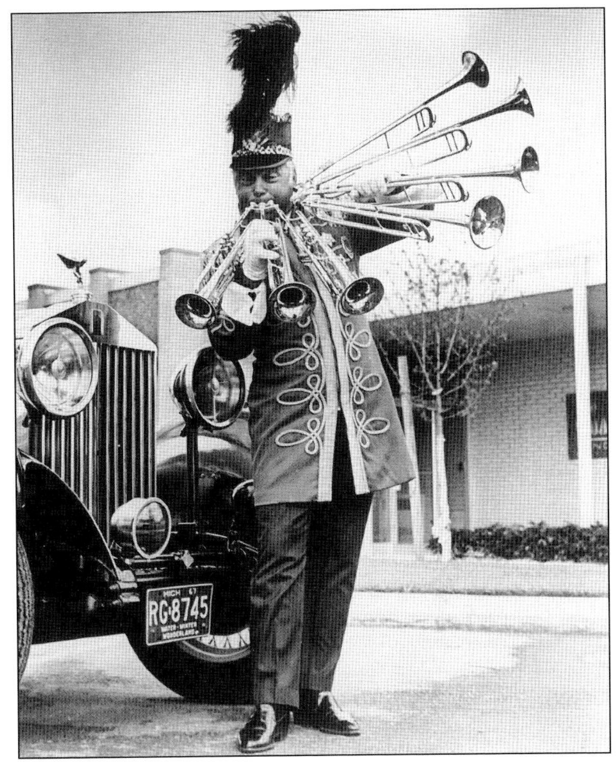

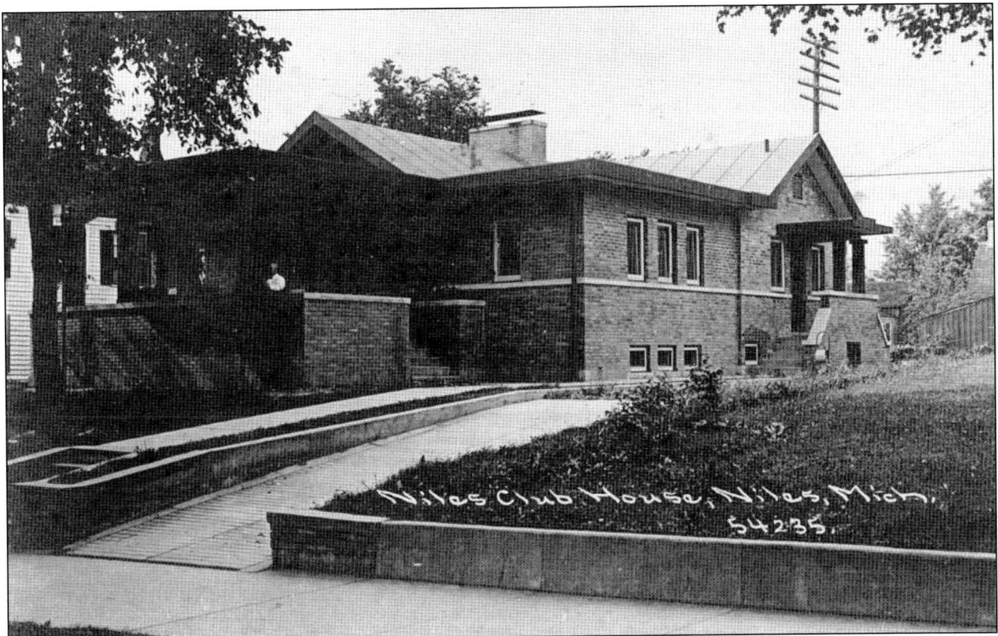

Walter Burley Griffin designed the unique Niles Club building in 1910 to house the Niles Club, a businessmen's organization. It was located next to the Carnegie Library on Main Street. Griffin, a Chicago-based architect, is credited with designing hundreds of buildings and completing the city plan for Canberra, the capital of Australia. Griffin trained under Frank Lloyd Wright and incorporated elements of Wright's Prairie School design.

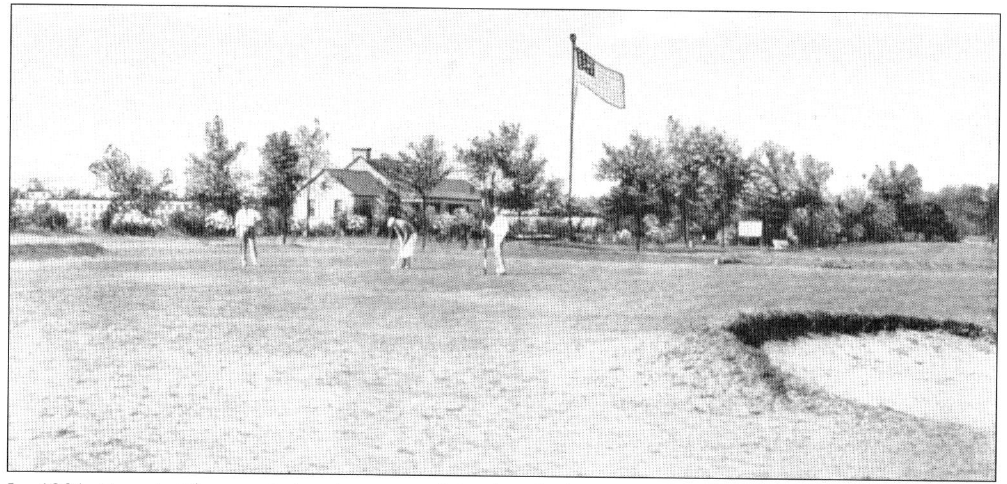

In 1921, Francis Plym and the city council finalized plans to turn 67 acres of land located near Kawneer's headquarters on Front Street into a municipal park. The park grounds included a baseball diamond, tennis courts, and a golf course with a clubhouse, shown here. To maintain the links, many Nilesites, including Ring Lardner, who was living in New York at the time, bought memberships to the Plym Park Golf Association.

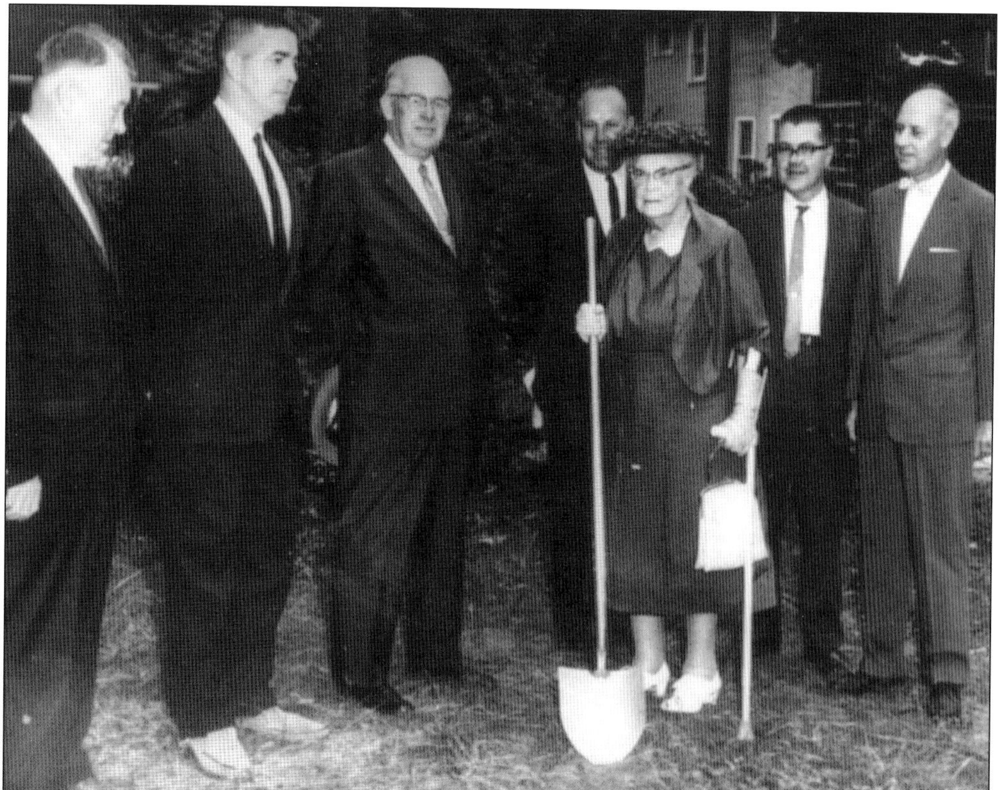

In 1962, Jennie Plym, of Kawneer fame, announced plans to fund a new public library building. Jennie attended the groundbreaking ceremony, posing for a photograph with library board members and staff. The new building was four times larger than the earlier Carnegie Library and featured modern design elements, including a glass rotunda. (Courtesy of the Niles District Library.)

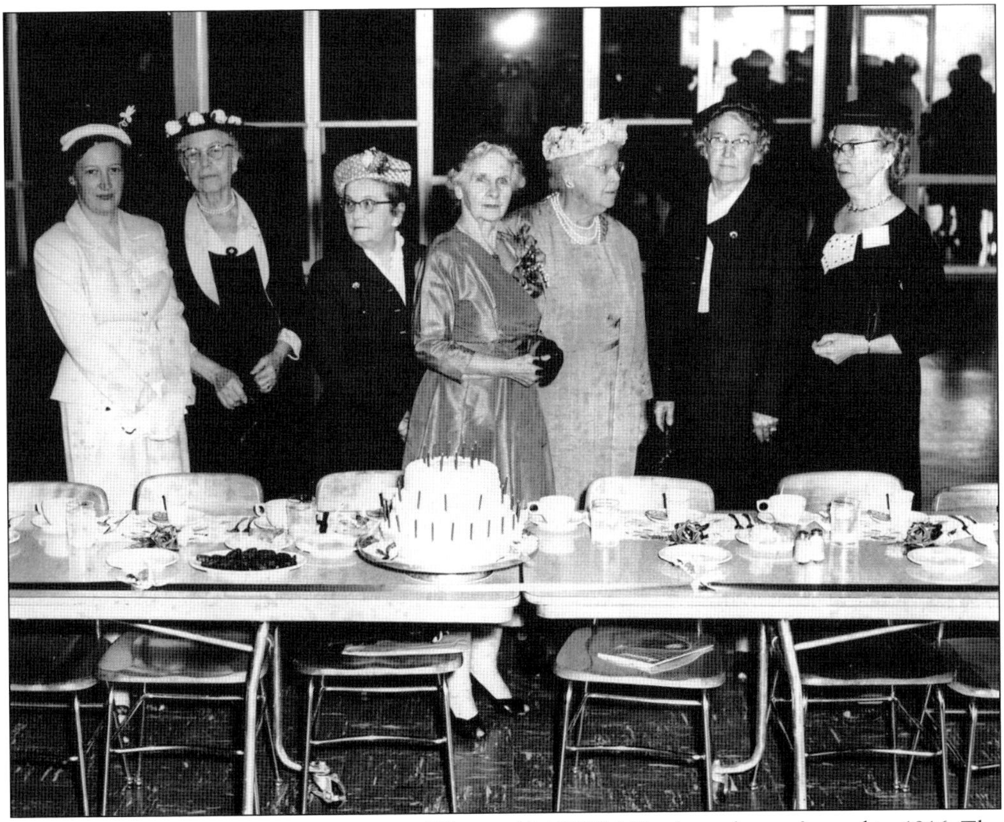

The American Association of University Women (AAUW) Niles branch was formed in 1916. The group promoted higher education for women and advocated for gender equality in the workplace. Past and present members of the AAUW Niles branch celebrated the club's 40th anniversary in 1957. Pictured from left to right are Marion Parkinson, Margaret Barber, Edith Hull, Helen Travis, Jennie Plym, Alta Wood, and Gladys Krenshaw.

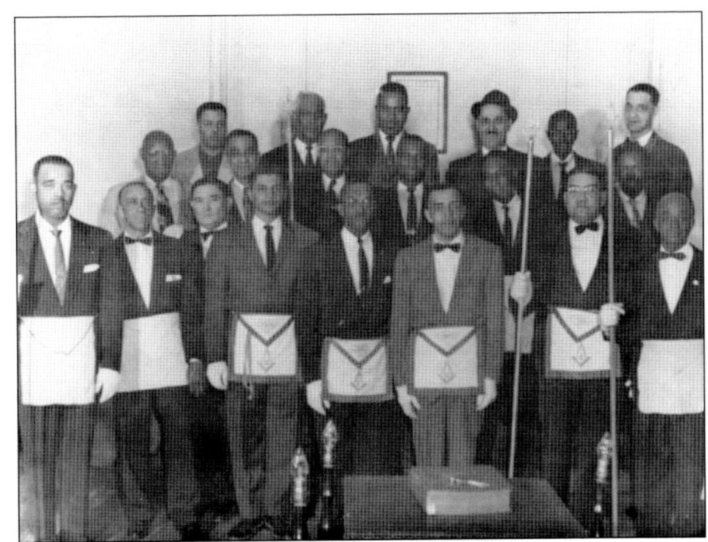

Early Niles leaders and businessmen belonged to Masonic organizations. Denied entry to white Masonic lodges, African American Prince Hall started his own organization known as Prince Hall Freemasons. Niles was the first place in Michigan with a Prince Hall Lodge. Harrison Lodge was organized in 1859. It later became the John W. Moore Lodge and met at Seventh and Ferry Streets. The group is pictured here in the 1960s.

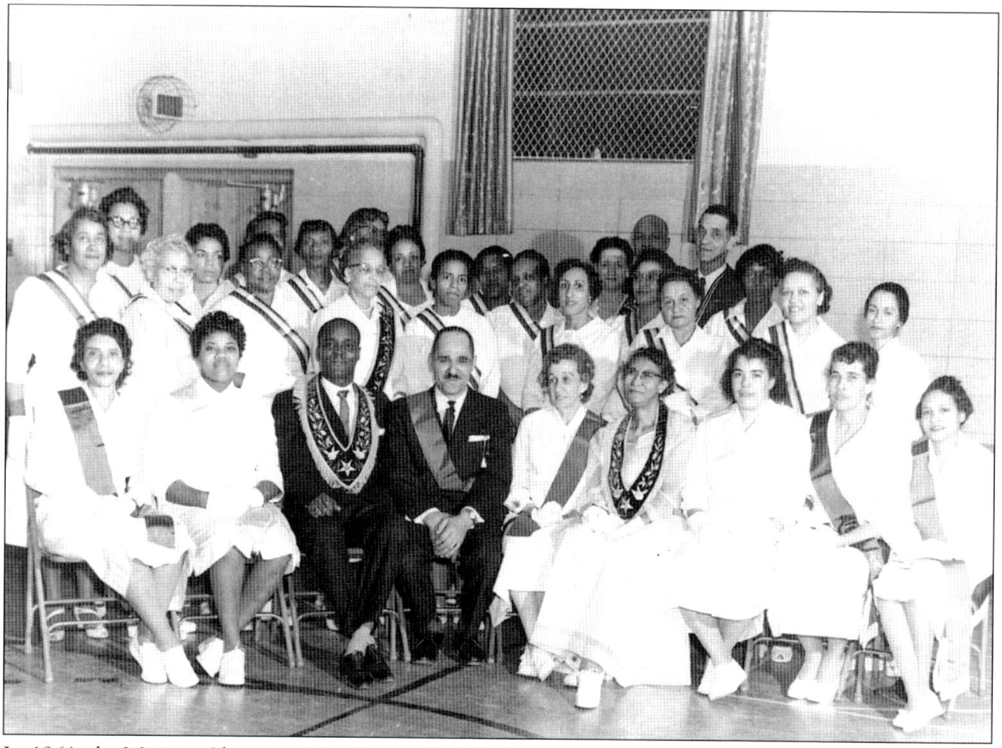

In 1961, the Miriam Chapter of the Prince Hall Order of the Eastern Star was established for African American women. At their first meeting at Northside School, Loren Anderson was elected Worthy Matron. Eastern Star members worked on charitable and service projects, including hosting penny suppers for the community and packing Red Cross Christmas "ditty bags" for servicemen overseas.

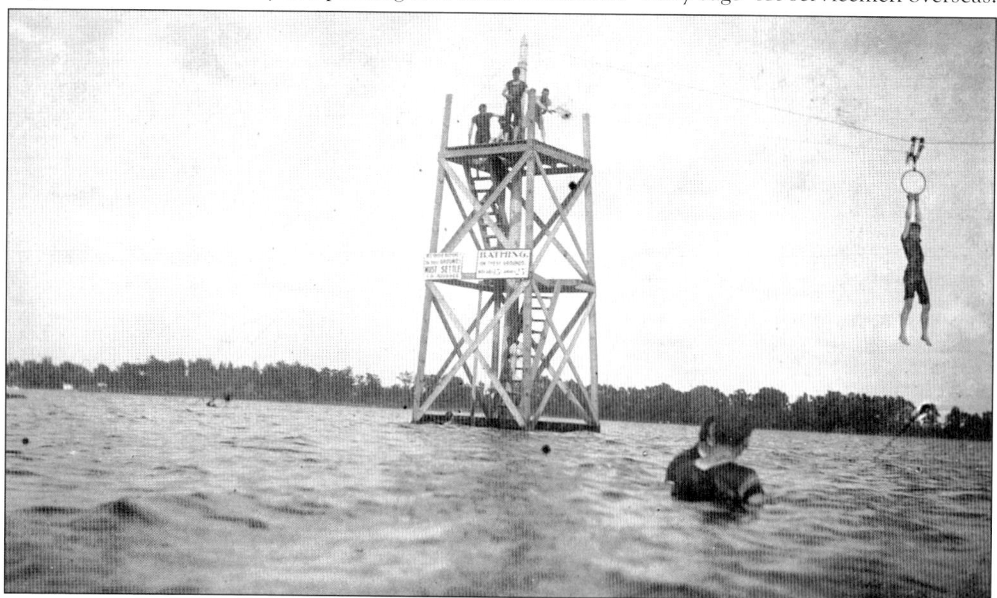

Located east of the city limits, Barron Lake was a popular recreation spot for Nilesites and regional tourists. The lake featured boating and water sports, with resorts and dance halls along the shore. The railroad made regular stops at Barron Lake and brought scores of tourists.

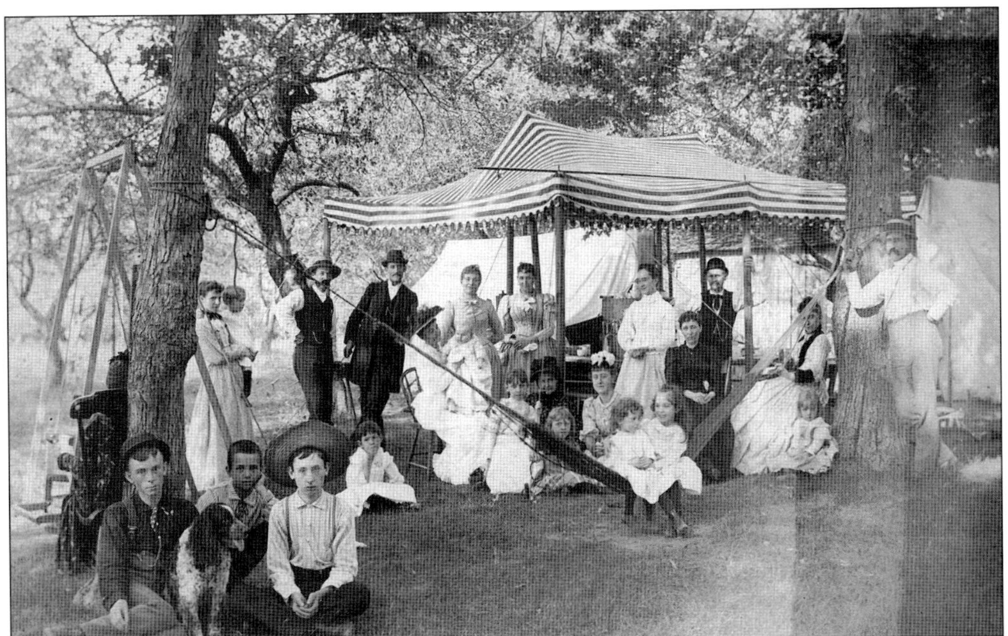

Members of a local family enjoy a picnic under a striped pavilion near Barron Lake. This photograph dates to around 1890. Barron Lake was a major draw since the early days of Niles. An 1850 article from *The Niles Democrat* mentions that the lake featured sandy bottoms, warm clear water, and beautiful shade on the border, which was being improved by hundreds of ladies working on landscaping.

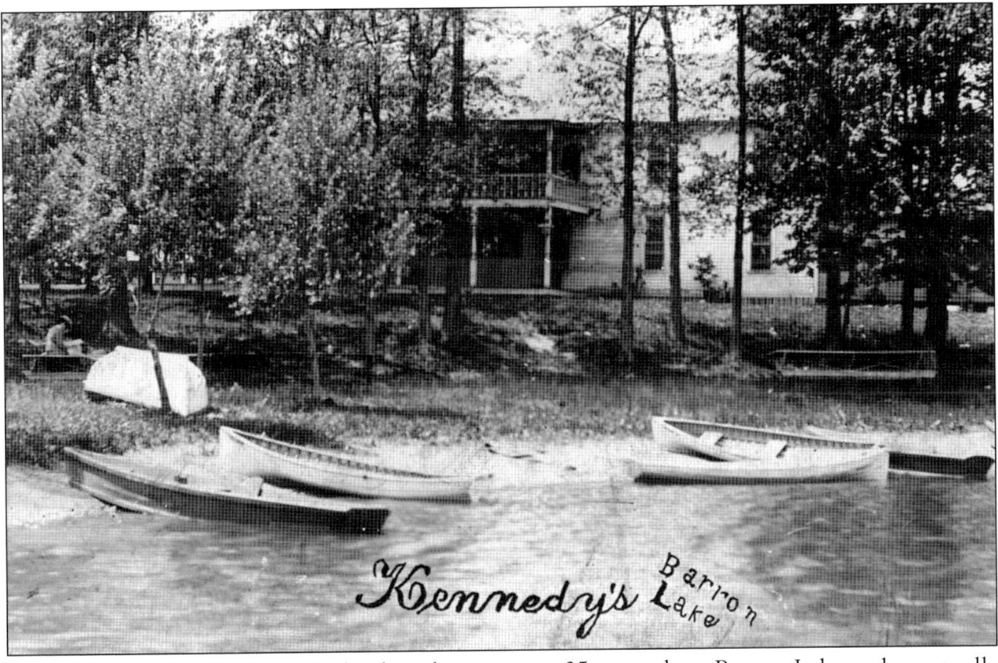

In the 1890s, Richard Kennedy developed a resort on 35 acres along Barron Lake and eventually built a dance hall that extended out over the water. Kennedy's Resort offered bathing, boating, fishing, dining, and dancing. Bathing suits were available for rental. The dance hall was converted into a roller rink during the 1940s.

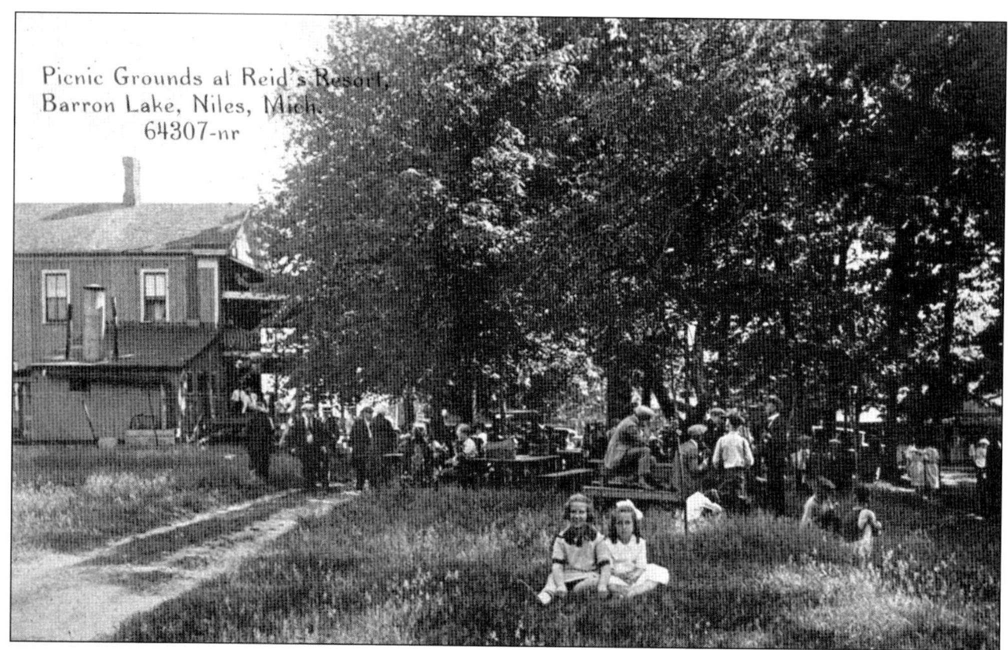

In 1911, William and Delia Reid moved to the eastern shore of Barron Lake and purchased the former Miller Resort along with eleven acres of land. Reid's Resort expanded over the years, and in 1935, a spectacular dance pavilion opened. The Reids often featured performer Don Pablo, whose performances were broadcast live by South Bend's WSBT radio. Players from the Chicago Cubs visited often. The resort was razed in 1949.

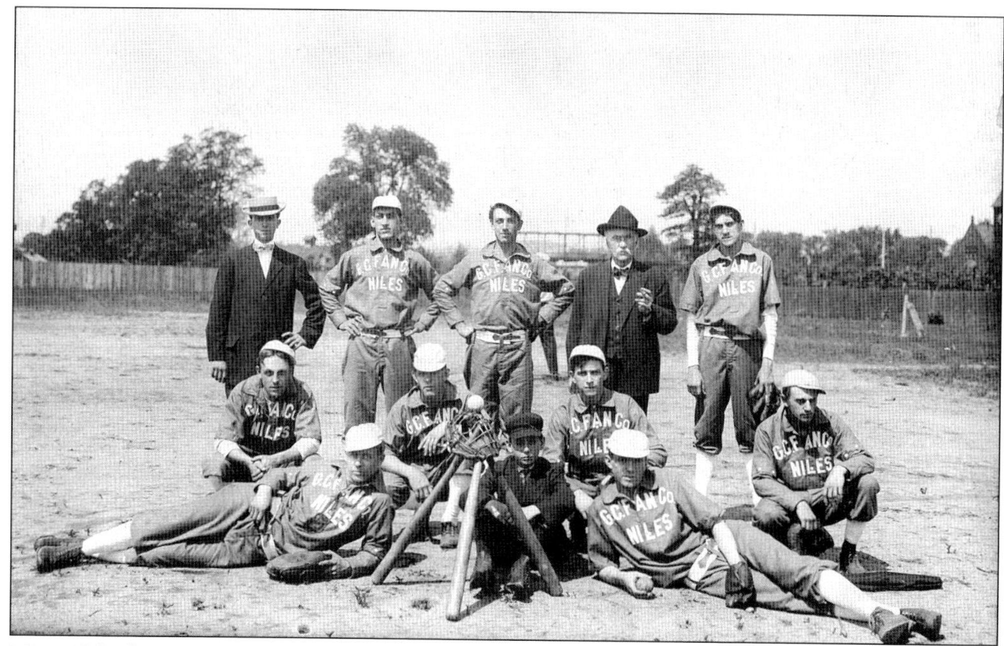

Many Niles businesses and industries sponsored extracurricular activities for employees and their families, including sports teams and clubs. Local industry ball leagues were popular throughout the 20th century. Shown here is an early 1900s photograph of the Garden City Fan Company baseball team. The Niles railroad depot and Fifth Street viaduct are visible in the background.

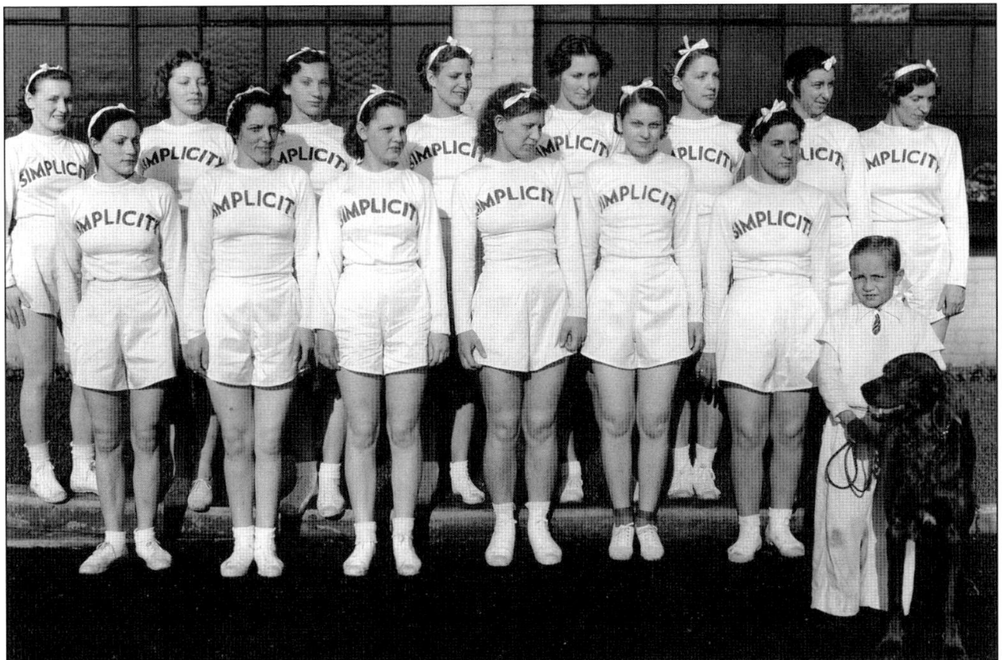

The 1935 Simplicity Pattern Company women's softball team was called the Himmys. Simplicity's *Impressions* newsletter notes that 1935 was the first year that the company participated in organized summer sports. The team donned uniforms with the company's adopted colors of cardinal and white. The local papers that year boasted that the Niles women's industrial league consisted of six teams, by far the most of any area town.

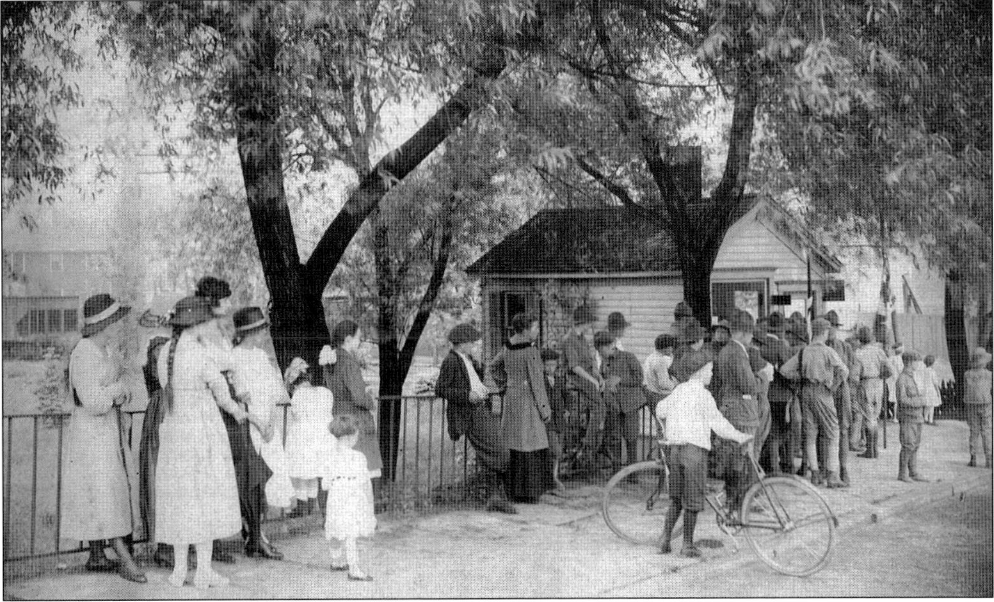

Onlookers gather near the Main Street bridge on May 30, 1918, for the Memorial Day parade. With World War I still raging, hundreds of families turned out to honor the deceased and those still serving overseas. Newspapers note it was the largest crowd in the city's history. Spanish-American War veteran George E. Corell served as parade marshal.

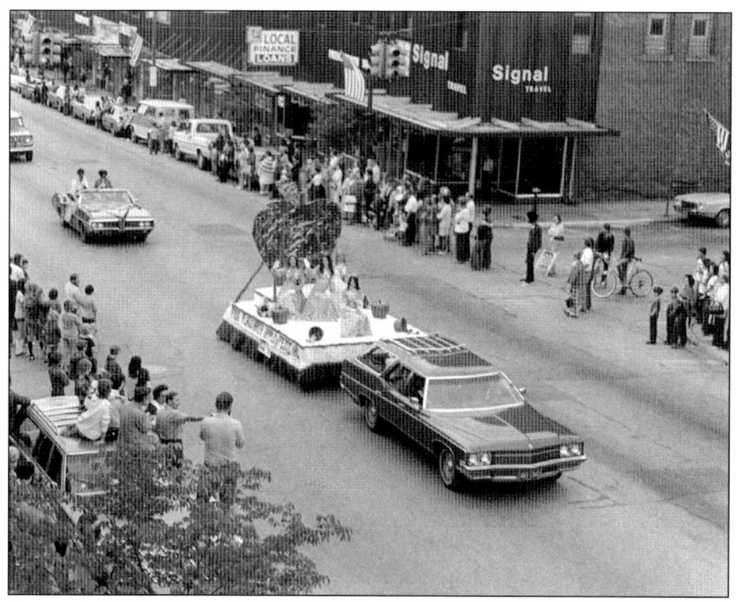

The Four Flags Apple Festival started in 1973 to celebrate the apple harvest and features carnival rides, an arts and crafts fair, food vendors, and more. The grand parade is a highlight of the popular festival. This photograph shows the Apple Festival royalty float going east on Main Street during the 1973 parade. Michigan governor William Milliken served as grand marshal of the parade that year.

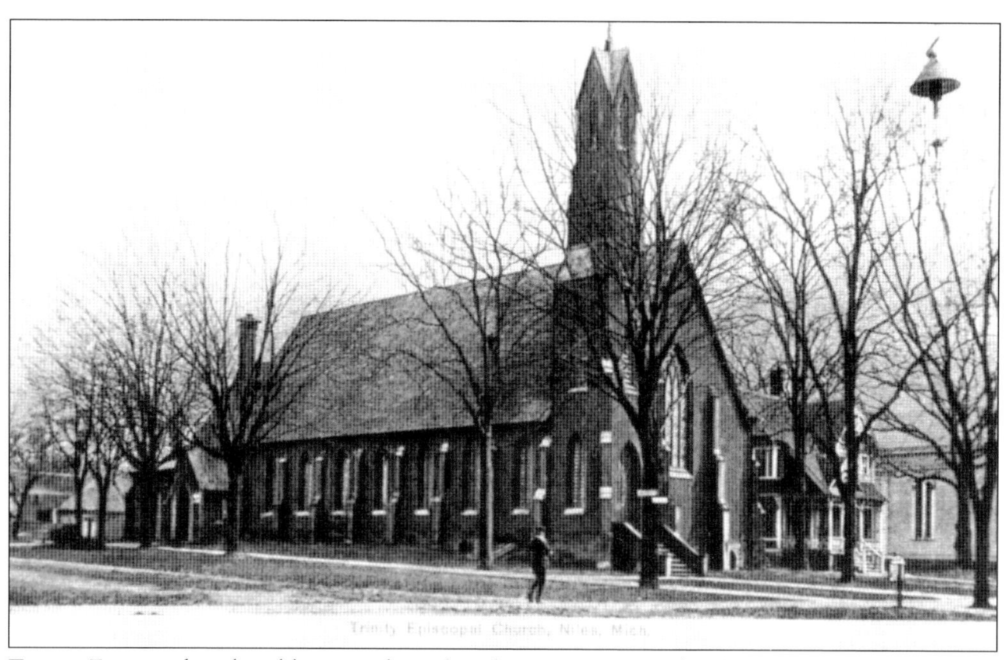

Trinity Episcopal is the oldest standing church structure in Niles, built in 1858. Rev. Joseph Phillips led the first service in the church on April 25, 1858. He was the grandfather of author Ring Lardner. Ring sang in the choir and played organ on occasion. His mother, Lena Bogardus Phillips Lardner, taught Sunday school and directed music. Her daughter (Ring's sister) Lena served as organist-choirmaster for over 40 years.

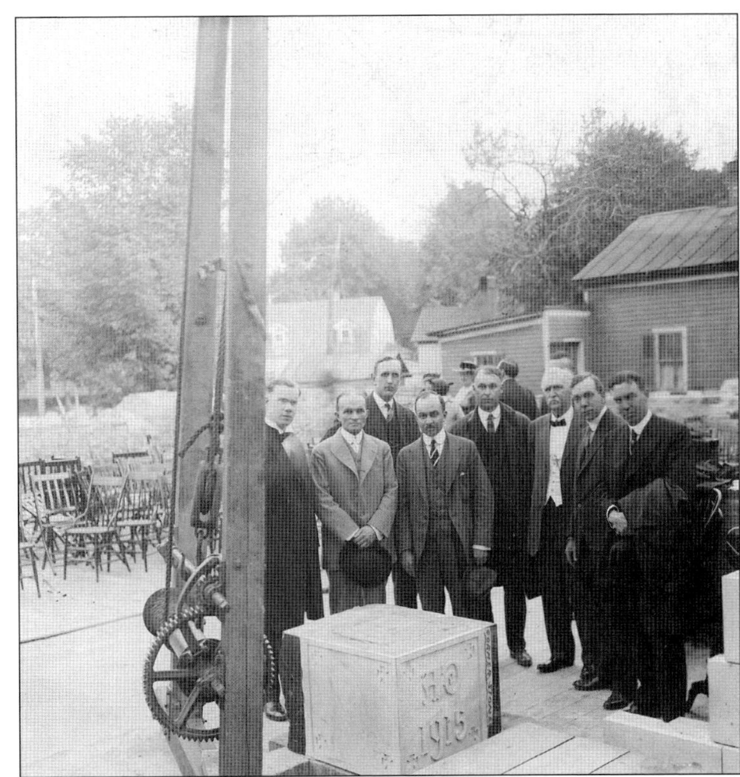

Building committee chairman Nathaniel H. Bacon, second from left in the first row, stands with members of the Chapin and Coolidge families at the cornerstone ceremony for Chapin Memorial Church. After the death of Charles Chapin in 1913, his will set aside $200,000 for philanthropic purposes. The Chapin estate funded various projects locally, including the building of a new Niles Presbyterian Church, which opened in January 1916.

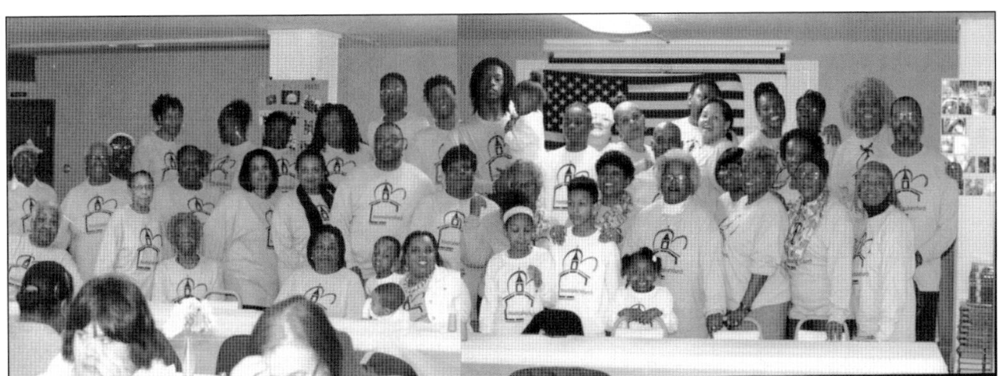

Second Baptist Church was organized in 1849 and was fully established by 1851 under the auspices of the Anti-Slavery Baptist Association. Rev. J.W. Hackley and D.G. Lett were the first pastors. The church was renamed Mount Calvary Baptist Church and still stands at Sixth and Ferry Streets. One of the oldest Black churches in Michigan, a historic site marker was placed here in 1994. This photograph celebrates the church's 170th anniversary celebration, which took place in 2019. (Courtesy of Mount Calvary Baptist Church.)

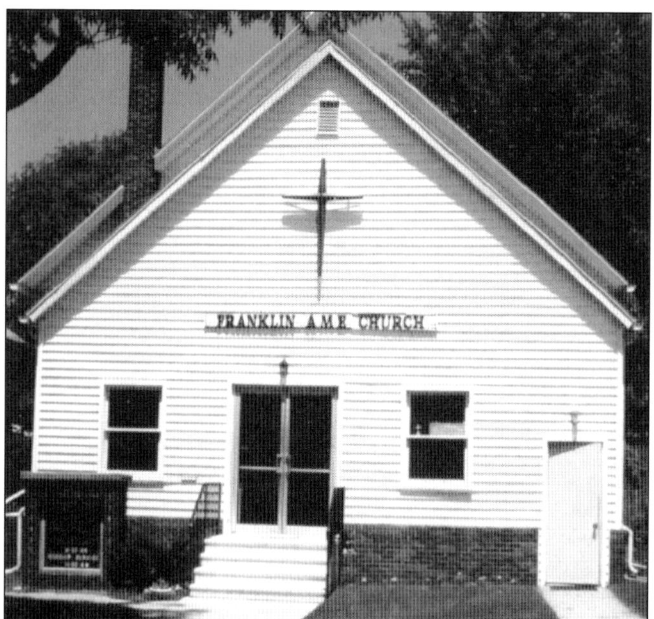

The Franklin African Methodist Episcopal Church has stood at Sycamore near Ninth Street since 1888. Rev. Cyrus Hill was the first pastor at this building, though the Methodist Society began meeting in Niles as early as 1860. In addition to hosting worship services, Franklin A.M.E. also serves as a community center. Notable civil rights activist Rosa Parks visited the church more than once.

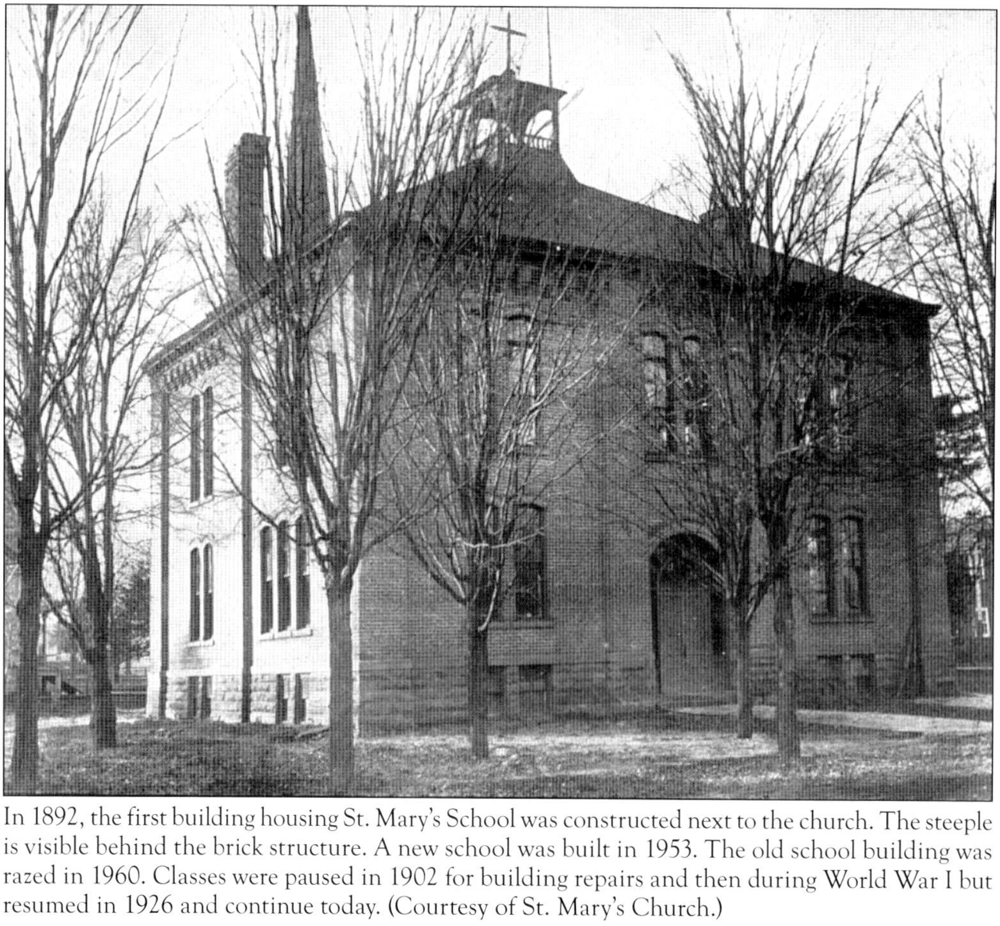

In 1892, the first building housing St. Mary's School was constructed next to the church. The steeple is visible behind the brick structure. A new school was built in 1953. The old school building was razed in 1960. Classes were paused in 1902 for building repairs and then during World War I but resumed in 1926 and continue today. (Courtesy of St. Mary's Church.)

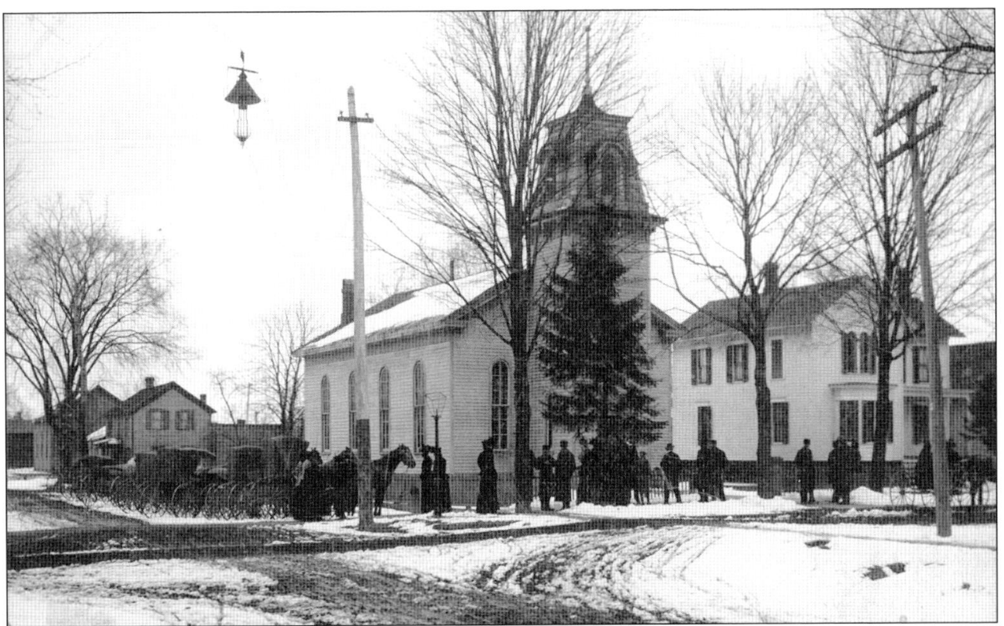

In 1858, German residents started holding church services and fellowship in a schoolhouse on the corners of Cass and Third Streets. As the congregation grew, a white frame structure, shown above, opened at the corner of Sixth and Sycamore Streets in 1865. Originally called St. John's German United Evangelical Church, the congregation conducted services in German until the early 1940s. A larger church was built in 1899.

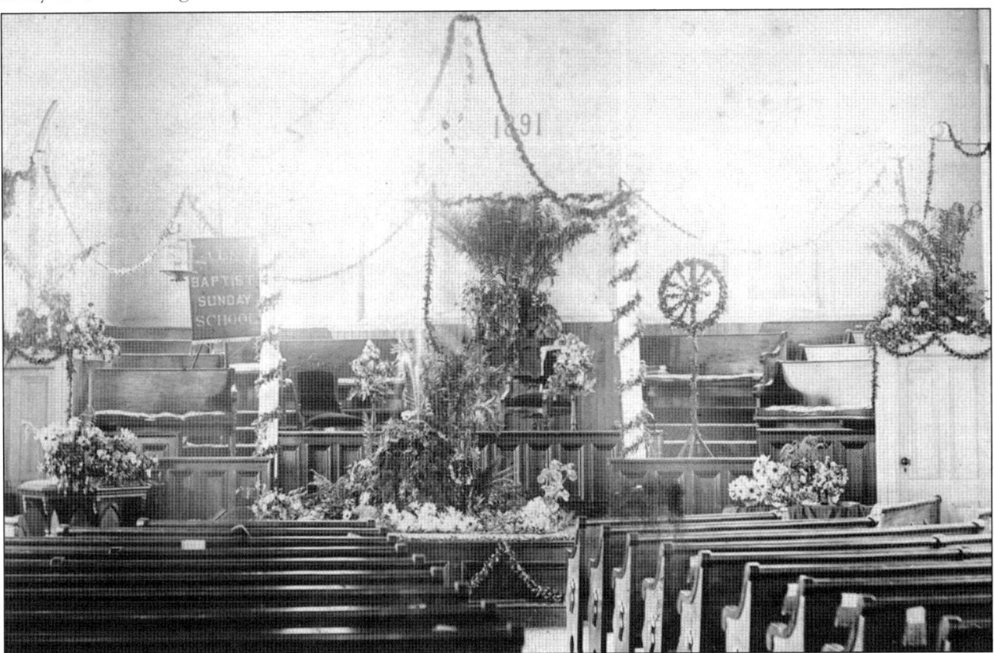

On the second Sunday in June, churches across the country observed Children's Day, an annual event organized by Sunday schools. At First Baptist Church, the 1891 Children's Day featured an abundance of floral arrangements on the altar. Many congregants took part in a responsive service, where Sunday school attendees contributed scripture readings and songs.

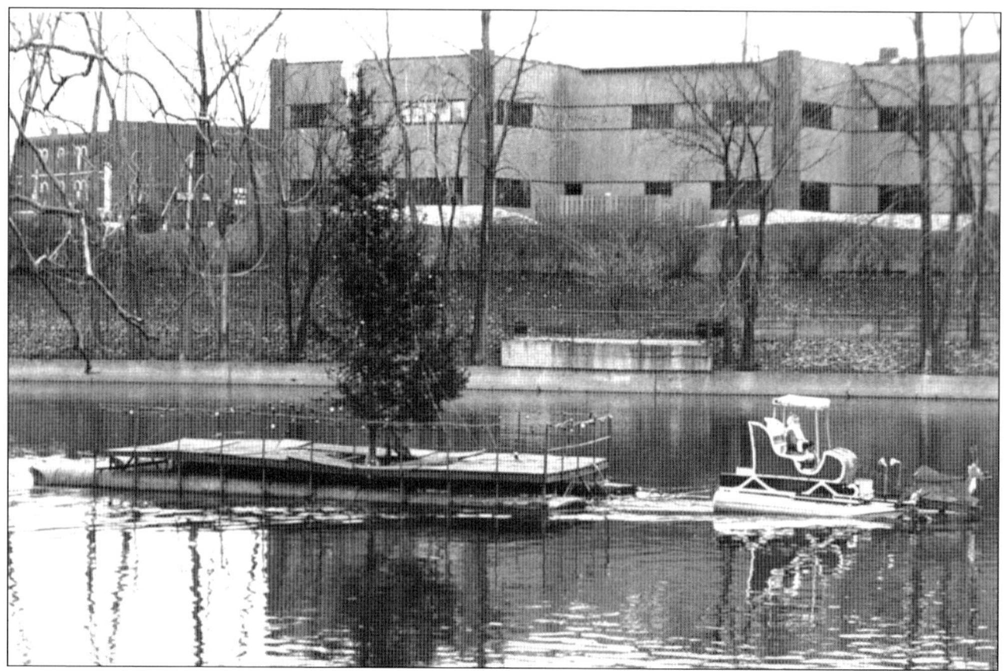

Roy Peters and several friends started the annual tradition of Santa in the River in 1962. At that time, the children's ward of Pawating Hospital overlooked the St. Joseph River. Peters had the idea for Santa to dock here and offer some holiday cheer. The sleigh lights up at night and includes reindeer, Christmas trees, and of course, the big man himself.

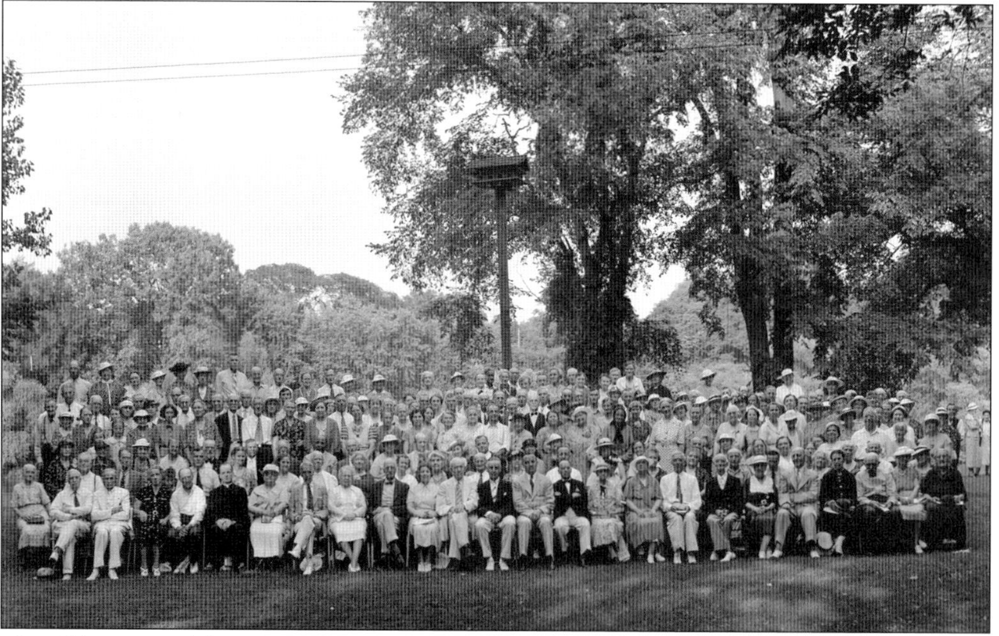

The Old Timers Club meets for a picnic at Island Park. Among the attendees at this 1937 event are Dr. Fred Bonine and his daughter Natalie Bonine French. Anna Babcock, age 94, was interviewed for being the oldest Nilesite present. Babcock was born in Kentucky, but her family moved here in 1862 because her father did not want to be drafted into the Confederate army.

About the Niles History Center

The Niles History Center (NHC), a division of the City of Niles's Community Development Department, is comprised of the Fort St. Joseph Museum and the Historic Chapin Mansion. Founded in 1939 at the former carriage house of the Chapin Mansion, the museum features exhibits that explore the story of Niles from the prehistoric era through the 17th-century founding of Fort St. Joseph to the modern business and industrial era. The Chapin Mansion is the former home of prominent citizens Henry and Ruby Chapin. Constructed in 1882, the Chapin Mansion was gifted to the City of Niles in 1933, served as the city hall until 2012, and is now administered by the history center.

The mission of the Niles History Center is to connect the past, present, and future. The Niles History Center advances its mission through educational programs, exhibits, and preservation of historical collections. The NHC, along with Western Michigan University, is an active partner in the Fort St. Joseph Archaeological Project and assists with the planning, research, and interpretation of the site.

In 2025, the Niles History Center was the recipient of two Governor's Awards for Historic Preservation: one for the rehabilitation of the Historic Chapin Mansion and one for the Fort St. Joseph Archaeological Project.